Observing Projects Using Starry Night Enthusiast™

For use with Freedman, Geller, and Kaufmann's

Universe

Ninth Edition

Marcel W. Bergman
T. Alan Clark
William J. F. Wilson

W. H. Freeman and Company
New York

© 2010 by W. H. Freeman and Company

ISBN-13: 978-1-4292-5886-9
ISBN-10: 1-4292-5886-1

Printed in the United States of America

First printing

W. H. Freeman and Company
41 Madison Avenue
New York, NY 10010
Houndmills, Basingstoke RG21 6XS England

www.whfreeman.com

CONTENTS

PREFACE

Astronomy is an observational science. The opportunity to make meaningful observations of the real sky is vital to the study of astronomy. However, for various reasons, such as light pollution or the lack of adequate equipment or re s o u rces, many students of astronomy are denied this opportunity. The projects described in this book provide an effective substitute for this observing experience by allowing students to make significant and accurate observations of the virtual sky using the *Starry Night Enthusiast™* program as well as hopefully enhancing the experience of students blessed with the re s o u rces to make real ground-based observations.

The "virtual" telescope provided by this package allows students to make observations that accurately convey current astronomical knowledge, even though these observations might not be possible to make in the real world. For example, observations that require months of real time to accumulate can be made with a condensed time animation. Observational hindrances such as daylight and the horizon can be removed to make measurements more easily understandable. The software allows phenomena and events that can be difficult to understand from the reference frame of the rotating and orbiting Earth to be viewed from a simpler and more elegant frame of reference. This change of perspective often makes obvious an otherwise subtle or difficult concept.

The projects in this book allow students to examine the appearance and behavior of various astronomical objects interactively. Physical laws and astronomical concepts become more interesting when students can demonstrate them with their own "observations." As examples, the observation of the 76-year orbit of Comet Halley underlines the crux of Kepler's second law, while the measurement of stellar parallax makes cosmic distances at once less daunting and more wondrous. The projects provide an initial framework for students and then guide them in their observations and the subsequent scientific analysis. Interspersed questions help students to focus their observations on particular concepts or theories. The beauty and wonder of the cosmos unfold throughout this sequence, as students make observations of the planets and their satellites, the nearby stars, deep-sky objects, and finally the three-dimensional soap-bubble structure in the realm of galaxies, galaxy clusters, and superclusters.

These observing projects present basic ideas at several levels. Most begin by introducing a basic idea in a simple way, using different viewing locations to illustrate this idea in the most effective manner. Then the simulations are extended to include measurements and simple analysis to demonstrate specific physical principles. Many of the projects use measurements that cannot be carried out in real life. Some examine the universe from locations that are inaccessible to us but that provide insight into the underlying science. Several chapters in the book contain mathematical derivations, but these derivations always culminate in simple formulae to which the observations and measurements can be applied. Thus, the point of the observations can be demonstrated without the necessity of a full understanding of the mathematical derivations. Several exercises require graphing of the measurements in order to demonstrate a particular concept and include the elementary consideration of possible errors in the measurements. This approach—the application of mathematical formulae, graphing, and error analysis to measured results—is designed to introduce students to the methodology of science.

Configuring *Starry Night Enthusiast*™ to illustrate the various concepts of astronomy can often be quite complex and time-consuming and can lead to annoying set-up problems and frustration. To avoid these difficulties, configuration files for these initial conditions have been prepared for all projects. These files have been placed on the CD along with the main program. They are readily accessed from *Starry Night Enthusiast*™ and provide immediate access to the correct view on the screen, thereby allowing students to focus more clearly on astronomy rather than on software interface techniques.

The first project is a comprehensive tutorial covering most parts of the interface to the program. It allows students to become proficient in operating the whole system and become confident in the alternative techniques for accomplishing various tasks, including setting up their own scenarios to demonstrate other concepts. In addition, a User's Guide is available along with this software so that students can refresh their knowledge of operating steps and procedures. Students should refer to this guide and complete the whole Tutorial project as a prelude to carrying out the projects and experiment with the many features of this program before attempting the individual projects.

While some of the projects in this book are relatively simple in concept and execution, others will appeal to more advanced students. Nevertheless, in all exercises, we have taken care to test and retest the setup files and the instruction sequences so that they are very complete. They should permit all students to carry out the tasks efficiently and reproducibly.

It is a pleasure to thank the staff of W. H. Freeman and Company for the opportunity to develop this book and for their careful and professional approach to it. We thank particularly Amy Thorne for her guidance, constant encouragement, and professionalism throughout this project. We are very grateful for the work of Jodi Isman and Laura McGinn, as they shepherded the book through the editing stages with skill and a constant concern for quality. We appreciate the work of Janet Bidwell of Tactical Graphics for her speedy and expert translation of the text into typeset manuscript. We also thank Pedro Braganca and David Whipps and their team of software designers for *Starry Night Enthusiast*™ at Simulation Curriculum Corp. for their help in streamlining the inclusion of the pre-configured set-up files in this project. The careful and thorough review of this book by James Dickinson of Clackamas Community College and the generous provision of this review to the authors is gratefully acknowledged. Colleagues at the University of Calgary have been a constant source of ideas and suggestions during the writing of the current and previous editions of this book. Above all, the authors express their gratitude and thanks to their wives, Marcia, Jean, and Dawn, respectively, for their support during the development of this book.

Tutorial 1

The observing projects in this book are designed specifically for the version of *Starry Night Enthusiast*™ that accompanies the textbook: *Universe,* 9th Edition, by Freedman, Geller, and Kaufmann. Most of the observing projects rely on files that are included with this particular version of the software. Although largely platform-independent, some features of *Starry Night Enthusiast*™ are accessed differently for Windows and Macintosh users. Also, as you will learn from this tutorial, there are sometimes several different ways to accomplish a task in *Starry Night Enthusiast*™. Different users of the program will prefer one method to another. One method might be more convenient in one context and a different method more efficient in another context. The instruction steps in the observing projects reflect the authors' preferences and experience with the *Starry Night Enthusiast*™ interface. You may prefer other ways to accomplish the same task.

As you work through the observing projects, there may be times when you cannot remember the keystrokes or menu selections required to accomplish an instruction step. In that case, you can look at the *Starry Night Enthusiast*™ manual available under the **Help** menu, or you can consult the **SkyGuide** pane in *Starry Night Enthusiast*™, which includes a guide to the features of the program and a step-by-step introduction to the basics of the interface.

The observing projects in this book are built around specific sequences of instructions that will guide your interaction with the *Starry Night Enthusiast*™ interface and help you to make interesting and accurate astronomical observations. This tutorial will introduce you to some of the features of *Starry Night Enthusiast*™ while also familiarizing you with the format of the observing projects in this book.

A. Starting Assumptions

You should be familiar with the following techniques as they apply to your operating system:

a. Standard mouse and keyboard techniques

b. Menu and toolbar navigation

c. File management, including opening, saving, and copying files

Starry Night Enthusiast™ should be properly installed and registered and you should know how to start the program.

The projects in this book include sequences of instructions that help you to make astronomical observations of the *Starry Night Enthusiast*™ sky. Instruction steps look like this:

1. Launch *Starry Night Enthusiast*™.

If this is the first time that *Starry Night Enthusiast*™ has run on your computer, you will need to register the software. Once you have registered, you can choose to download data updates for satellites, comets, and asteroids. Following these housekeeping duties, *Starry Night Enthusiast*™ will ask you to set up your home location.

2. In the **Home Location** dialog box, click the **List** tab. Use the scrollbar to the right of the list to search for the name of your city or town in the list. Alternatively, type the name of your home location to jump to it in the list or, in North America, enter your Zip or Postal Code. Once you have found the name of your city or town, highlight it with a mouse click and then click the **Set Home Location** button. If you ever need to change your home location, select **Set Home Location...** from the **File** menu on a Windows system or from the **Starry Night Enthusiast** menu on a Macintosh.

With your home location set, *Starry Night Enthusiast*™ will open the main view window with a view of a southern horizon and a depiction of the sky as it would appear from your home location at the time and date specified by the system clock on your computer.

3. Verify that you have the correct version of the software by selecting **Help > About Starry Night Enthusiast...** from the menu. The version number is shown above your registration information in the lower left corner of the window that pops up. Click the **OK** button to close this window and return to the *Starry Night Enthusiast*™ screen.

There are four main sections in the *Starry Night Enthusiast*™ screen:
1) The main view window occupies the largest area of the *Starry Night Enthusiast*™ screen and depicts a view consistent with the parameters displayed in the toolbar.
2) The toolbar contains a collection of controls and displays, arranged across the screen above the main view window, which specifies parameters of the view such as the date, time, viewing location, and gaze direction.
3) The main menu is immediately above the toolbar.
4) Side-pane tabs, arranged vertically along the left border of the main view window, open various side panes that allow you to change options and other settings of the current view.

B. The Starry Night Enthusiast™ Toolbar

The *Starry Night Enthusiast*™ toolbar stretches across the screen directly above the main view window. It is divided into five sections labeled, from left to right: 1) **Time and Date**, 2) **Time Flow Rate**, 3) **Viewing Location**, 4) **Gaze**, and 5) **Zoom**. Below each section label, a panel displays the values appropriate to the section that are in effect in the current view. Below the display panel, each control section has a collection of buttons that allow you to vary the parameters for that section and thereby affect the view in the main window.

C. Time and Date Controls

The **Time and Date** control section of the toolbar shows the time and date of the current view. A small Sun icon in the upper left corner of the display panel indicates whether daylight saving time is in effect.

4. Click the **Daylight Saving Time** control several times. Each click toggles the effect of daylight saving time on or off. When the icon is bright, daylight saving time is in effect in the current view and the time in the display panel advances by 1 hour. When the icon is dull, daylight saving time is off and the time in the display moves back 1 hour.

There are several methods for changing the time and date. The next instruction sequence allows you to explore these methods.

5. Click the **Hour** field of the **Time and Date** section of the toolbar. The hour digit becomes highlighted. Now you can enter appropriate values for the hour directly from the keyboard. You will see the **Hour** value in the **Time and Date** display panel change as you type. You will also note that the view in the main view window changes to reflect the new time.

6. You can also increment or decrement the hour value in steps of 1 hour. With the **Hour** digit still highlighted, press the + key on the keyboard. Notice that the time advances by 1 hour. To move back 1 hour in time, press the — key on the keyboard. Holding down either key allows you to advance or retard the time rapidly, 1 hour at a time.

7. The techniques outlined in the previous two steps work with the **Hour, Minute, Second, Month, Date,** and **Year** fields in the **Time and Date** display. Once you have selected any one of these fields you can quickly navigate back and forth through all of these fields by using **Shift-Tab** and **Tab**, respectively. Experiment by navigating through the time and date fields and by using different methods to change their values. Watch the view update to reflect the changes that you make.

8. Click on the AM/PM and BC/AD controls in the **Time and Date** display panel. Note that these fields simply toggle from one value to the other with each click and cannot be highlighted.

9. Double-clicking any numerical field or the **Month** field in the **Time and Date** display changes the time and date to that of the system clock of your computer. After double-clicking the **Time and Date** display, none of its fields should be highlighted.

10. The **Hour, Minute, Day, Month,** and **Year** fields of the **Time and Date** display can be incremented and decremented by using specific shortcut keys. To increment the **Hour** field press the H key on your keyboard. Holding the key down causes time to advance rapidly in increments of one hour. To decrement the hour, type **Shift-H.** The shortcut keys for incrementing and decrementing **Time and Date** fields are shown below:

Field	Increment	Decrement
Hour	H	Shift-H
Minute	T	Shift-T
Day	D	Shift-D
Month	M	Shift-M
Year	Y	Shift-Y

11. At the far right of the **Time and Date** display panel, a downward-pointing arrowhead gives you access to times of astronomical significance. Click this arrow and select from the list of times in the drop-down menu.

12. The three buttons at the bottom of the **Time and Date** control section allow you to change the time and date quickly to **Now** (the time and date specified in the system clock of your computer) or to **Sunrise** or **Sunset** on the date shown in the display panel.

D. Time Flow Controls

To the right of the **Time and Date** section of the toolbar is the **Time Flow Rate** control section. This section of the toolbar consists of a display panel that shows the current setting for the rate at which time flows in the view when activated by the buttons below the display, which look like typical CD or video control buttons. When *Starry Night Enthusiast*™ first opens, the **Time Flow Rate** is set to the default value, **1×** and the **Run time forward** ("Play") button is highlighted in blue. This indicates that time is flowing in the view in the main window at the same rate as real time. You can verify this by studying the **Time and Date** display for a few moments.

13. Click on the number in the **Time Flow Rate** display to highlight it. With both the number in the **Time Flow Rate** display panel and the **Run time forward** button highlighted, type a number on the keyboard of up to three digits (say, 800). Watch the view and the **Time and Date** display change under the accelerated time flow, in this case 800 times faster than real time. With the number in the **Time Flow Rate** display highlighted, you can increment or decrement this value using the **+** and **−** keys on the keyboard.

14. Click on the icon of a downward-pointing arrow to the right of the value in the **Time Flow Rate** display panel. This brings up a menu that contains two groups of options separated by a horizontal line. Selecting a menu item in the group of options above this line allows you to set the **Time Flow Rate** to a specific multiple of real time. Select **3000×** from the menu and watch the view. Note that, when a time flow rate is selected from the group above the dividing line in the menu, the change takes place immediately, with time flowing in the view. When you select an item from this menu from the group of interval options below the dividing line, the **Time Flow Rate** is set to one unit of the interval and the **Stop time** button in the **Time Flow Rate** controls is highlighted in red to indicate that time flow in the view has ceased. Now you can choose any multiple of the time interval that you selected from the menu by clicking on the numerical value in the display and typing a specific number of intervals from the keyboard or by using the **+** and **−** keys to increment or decrement this value.

15. Open the **Time Flow Rate** menu and select **days.** Then click the **Run time forward** button and watch the view and the **Time and Date** display change as time flows.

16. Click the **Run time backward** button (the button to the immediate left of the **Stop** button with the icon of an arrow pointing to the left) and watch as the direction of time flow reverses and time changes smoothly in intervals of 1 day.

17. Click the **Step time backward** button (the leftmost button of the time flow controls) as you watch the **Time and Date** display. Notice that each time you click the **Step time backward** button, the date in the display decreases by 1 day. Clicking the **Step time forward** button (the rightmost button of the time flow controls) does the opposite. Each click of the **Step time forward** button advances the time by 1 day. You can also use the keyboard shortcuts U and **Shift-U** to **Step time forward** or back, respectively, in units of time specified in the **Time Flow Rate** display panel.

18. Press the R key on the keyboard. This resets the time to **Now** and the rate of time flow to **1×** (real time). Press **Shift-R** to reset the **Time Flow Rate** to real time (**1×**) without resetting the time to **Now.**

19. Experiment with the **Time Flow Rate** controls to determine the earliest and latest dates that *Starry Night Enthusiast™* can display.

Interspersed through the observing projects are questions that help to ensure that you understand the observations that you make.

> **Question 1.** What is the earliest date that you can display in *Starry Night Enthusiast™*?

> **Question 2.** What is the furthest date in the future that you can display in *Starry Night Enthusiast™*?

E. Viewing Location Controls

The center section of the toolbar contains a panel displaying your current **Viewing Location.** When you start *Starry Night Enthusiast™*, this panel should show the name of the city or town that you specified as your home location. Below the **Viewing Location** display panel are four buttons. The two buttons on the left, with arrowhead icons pointing down and up, are elevation controls. These controls allow you to change elevation over the current viewing location.

20. Repeatedly click the **Increase current elevation** button (up arrowhead icon) as you watch the main view and the **Viewing Location** display. With each click of this button, your elevation above your home location on Earth increases and this is reflected in the view window and in the **Viewing Location** display panel.

21. With the mouse cursor positioned over the **Increase current elevation** control, hold down the mouse button (the left mouse button on a Windows system) and watch as you lift off rapidly from the surface of Earth and move into outer space.

22. Click the **Decrease current elevation** button (down arrowhead icon) to observe the effect of this control. Notice that the **Decrease current elevation** control does not allow you to descend below the planet's surface.

23. Click the **Spaceship** button below the **Viewing Location** display. The view alters as you soar into space and the **Viewing Location** display panel changes to read "Spaceship." In the upper left corner of the view window are instructions for using the keyboard to control your space flight. Experiment with these controls and have fun flying through the universe at warp speeds!

24. When you are done experimenting with the spaceship, click the **Spaceship** button to exit spaceship mode.

25. Click the **Home** button to return to your home location on the surface of Earth. Note that the **Home** button not only returns you to your home location but also resets the time and date to **Now** and the time flow to real time. You can also use a keyboard shortcut, **Ctrl-Shift-H** (**Cmd-Shift-H** on a Macintosh) to go **Home**.

F. Gaze Direction Controls

To the right of the **Viewing Location** section of the toolbar is the **Gaze** section. The **Gaze** section consists of a display panel indicating the altitude and azimuth of the gaze direction for the view. When *Starry Night Enthusiast*™ first opens, or after clicking the **Home** button in the **Viewing Location** section of the toolbar, the view is of a southern horizon (gaze azimuth 180°) and part of the southern sky (gaze altitude of 25°). In addition to the display panel, the **Gaze** section of the toolbar has five buttons labelled **N, S, E, W,** and **Z.**

26. Click the **N** button. The view slowly pans across the screen and stops when the gaze direction has shifted toward due north (azimuth 0°). As you may have guessed, clicking the **S, E,** and **W** buttons changes the gaze direction to south, east, and west, respectively. Instead of clicking the buttons in the **Gaze** section of the toolbar, you can accomplish the same effect by typing the first letter of a direction on the keyboard. For example, type **E** on the keyboard to change the view direction to due east (azimuth 90°).

27. The **Z** button in the **Gaze** section of the toolbar changes the gaze direction to your zenith, straight up from your viewing location (altitude 90°). Click the **Z** button or type the letter **Z** on the keyboard to see this effect.

28. To change the gaze direction in smaller steps, you can use the cursor keys on the keyboard.

29. Another way to change the gaze direction of the view is to use scrollbars. Select **Show Scrollbars** in the **View** menu. Scrollbars appear along the right and bottom border of the main view window. You can use standard mouse techniques to move the scrollbars in order to change the gaze direction. When you are done experimenting with the scrollbars, select **View > Hide Scrollbars** in the menu.

You will note that, when using the **Gaze** controls to change the view, a temporary display appears in the upper right-hand corner of the window, showing the current field of view and view direction. This display is useful in providing orientation as you scan the sky.

G. Zoom Controls

The final section of the toolbar, at the far right, is the **Zoom** section. This section includes a display pane that indicates the current width and height of the field of view. Below this display panel are the two zoom buttons.

30. Change the **Time** in the **Time and Date** panel to **12:00:00** AM (daylight saving time off). Then watch the view and the **Zoom** display while repeatedly clicking, or clicking and holding, the **Zoom In** button (the button with the plus sign). Next, repeatedly click or click and hold the **Zoom Out** button (the button with the minus sign) to see how these controls work. You can also use the keyboard to zoom in (press the **+** key) and zoom out (press the **–** key). As you zoom in and out, the **Zoom** display panel shows you the width and height of the current field of view. As you change the field of view, the temporary display in the upper right-hand corner of the view shows this field diagrammatically.

Question 3. Approximately what is the minimum field of view in *Starry Night Enthusiast*™?

Question 4. What is the maximum width of the field of view in *Starry Night Enthusiast*™?

H. Grabbing Hold of The Main View Window

The main view window will be your observatory. By using the controls discussed above to change the parameters of the view, you will be able to make meaningful observations of objects and events in the sky.

31. Click the **Home** button in the toolbar to return to your home location.
32. **Stop time.**
33. Change the **Time** to **12:00:00** PM.
34. Without clicking any mouse buttons, move the mouse cursor around over the daytime sky in the view. Notice that the cursor is in the shape of a small hand. This cursor is called the **Hand Tool.**
35. With the **Hand Tool** positioned over the sky in the view, click the mouse button (the left mouse button on a Windows system) and you will see the cursor change into a fist. While pressing the mouse button, drag the mouse and observe what happens to the view. Also watch how the **Gaze** panel display changes to reflect your movements.

While you can certainly use the controls discussed earlier to shift the gaze direction, using the **Hand Tool** is often much simpler.

36. Click the **Home** button.
37. **Stop time.**
38. Change the **Time** to **12:00:00** AM.
39. Without clicking a mouse button, move the cursor around in the view, pausing over several stars or other objects in the sky. You will notice that, when the cursor rests over an object in the sky, its shape changes to a **Selection Tool** (an icon of an arrow with an accompanying list). In addition, information about the object over which the cursor is resting pops up on the screen. The information presented in this heads-up display, or **HUD**, can be customized, as you will learn later in this tutorial. Pressing the **I** key on the keyboard toggles the **HUD** on and off.
40. With the cursor resting over an object, and shaped like a **Selection Tool**, click the mouse button (the left mouse button on a Windows system) to **Select** the object, at which point a label appears next to the selected object.

41. Change the **Time** to **12:00:00 PM.**

42. Set the **Gaze** to North.

43. **Increase current elevation** to approximately **10,000 km** from Earth.

44. Change the **Gaze** direction to the **Zenith.** Position the mouse cursor near the top center of the view and click the mouse button to activate the **Hand Tool.** A small circular icon representing the zenith, or the point directly overhead at **Altitude 90°** appears near the center of the view. Using the **Hand Tool,** positioned near the top center of the view, click the mouse button and attempt to drag the view downward (i.e., try to look further up). Note that the icon marking the zenith stays fixed but that as you drag the **Hand Tool** to either side of the vertical centerline, the view rotates around the zenith point and the **Gaze** direction display panel varies only in **Azimuth.**

45. Change the **Gaze** direction to North.

46. Use the **Cursor Down** key to look down (or use the **Hand Tool** to drag the view upward) to change the **Gaze** direction so that the view shows the Earth near the center of the screen (the **Altitude** value in the **Gaze** display panel should be approximately −85°).

47. Position the mouse cursor near the top center of the view and activate the **Hand Tool** (click and hold the mouse button). Note that a small red icon (identical to the zenith marker that you encountered in a previous step) appears on the image of the Earth. This icon marks the **Nadir** point, the point looking directly downward from your location in space. Position the **Hand Tool** near the bottom center of the view over the space surrounding the Earth. Activate the **Hand Tool** and try to drag the view upward. Notice that the nadir point remains at a fixed location on the Earth. Do you recognize the geography of the Nadir point?

48. If necessary, use the **Hand Tool** to drag the view so that the image of the Earth is once more in the center of the view.

49. Deactivate the **Hand Tool** (release the mouse button) and position the mouse cursor over the image of the Earth. Notice that the cursor icon changes to a small square surrounded by four small arrowheads. This is the **Location Scroller.**

50. Position the **Location Scroller** near a limb (edge) of the Earth. Click the mouse button (the left mouse button on a Windows system) and drag the view. Notice that this produces a very different effect on the view from that of the **Hand Tool.** When you use the **Hand Tool** to drag the view, it is as if you remain in one stationary location and are looking around by turning your head in various directions. When you drag the view with the **Location Scroller,** it is as though you were flying around the planet or object centered in the view at a steady elevation. Use the **Location Scroller** over the image of the Earth to get used to this effect.

I. The Object Contextual Menu

In addition to the standard menu bar across the top of the application window, *Starry Night Enthusiast*™ offers contextual menus, so named because the options they contain vary with the context in which they appear. To activate these contextual menus, you must click the right mouse button on a Windows system whereas on a Macintosh system you must click the mouse button while holding down the **Ctrl** key on the keyboard.

51. Click **Home.**

52. **Stop time** and then change the **Time** to **12:00:00 PM.**

53. Position the mouse cursor over the blank sky. **Right-click** (Ctrl–Click on a Macintosh system) to invoke the contextual menu for that region of the sky. Examine the available menu options.

54. Select **Centre.** Invoking this command centers the selected region of the sky in the view. Dragging the view with the **Hand Tool** or moving the view with the cursor keys or scroll bars will break this lock of the view on that region of sky.

55. Select **Hide Daylight**. This option is self-explanatory. To turn daylight back on, select **Show Daylight**. (This command also appears under the **View** menu.)

56. Select **Show Ecliptic**. A green line that represents a projection of the Earth's orbit around the Sun appears across the view. Select **Hide Ecliptic** to toggle this display off.

When the contextual menu is invoked over an object (i.e., when the selection tool icon is active), different menu options will appear.

57. Click **Home**.

58. **Stop time**.

59. **Increase current elevation** to approximately **24,000 km from Earth**.

60. Set the **Date** to **June 30, 2007**.

61. Use the **Hand Tool** and the **Location Scroller** to adjust the view so that both the Moon and the Earth are visible.

62. Position the cursor over the image of the Earth and open the object contextual menu for the Earth by clicking the right mouse button (Windows) or by clicking the mouse while holding down the **Ctrl** key (Macintosh). Examine the options available.

63. Select **Centre**. The Earth is centered in the view.

64. Open the **Object Contextual Menu** for the **Moon** and select **Centre**. The view will now be centered upon the Moon.

65. Open the contextual menu for the **Moon** again and select **Orbit**. A line representing the Moon's orbit around the Earth appears in the view. Select **Orbit** again to turn this display off.

66. Select **Show Info**. The **Info** pane, containing detailed information on the Moon, opens to the left of the main view. Click the **Info** tab to close this side pane.

J. Labeling, Finding, and Centering Objects in the View

While you have learned to center objects that you see in the view by selecting **Centre** from a contextual menu, finding a particular object can be a frustrating process. *Starry Night Enthusiast*™ offers two solutions. The first solution is to have *Starry Night Enthusiast*™ label constellations or objects in the sky through the **Labels** menu in the main menu bar at the top of the screen.

67. Click **Home**.

68. **Stop time**.

69. Change the **Time** to **12:00:00 AM**.

70. Select **Labels > Show All Labels**. Labels to prominent objects and constellations appear in the view. The clutter can be somewhat bewildering, however. Select **Labels > Hide All Labels** to turn the labels in the view off again.

71. Select **Labels > Planets-Moons**. All of the planets that are visible from your location at the time and date displayed in the toolbar are labeled in the view. Navigate your way around the sky with the **Hand Tool** and make a list of those planets that are visible in your sky.

72. Experiment with the other options available under the **Labels** menu. Notice that all of these options toggle on or off each time that you select them.

While labels are handy, if you know the name of the object or constellation you are looking for, it is more convenient to find it directly.

73. Select **Edit > Find...** from the menu bar to open the **Find** pane. Alternatively, you can click the **Find** tab on the left of the main view window or use the keyboard shortcut **Ctrl-F** (**Cmd-F** on a Macintosh system) to open the **Find** pane.

The **Find** pane contains an edit box at the top that has a magnifying glass icon in its left margin. Below this edit box is a label describing the database that *Starry Night Enthusiast*™ will search when you type an object's name in the edit box. By default, *Starry Night Enthusiast*™ searches all of its databases.

74. Click the magnifying glass icon in the edit box to see a menu of the various *Starry Night Enthusiast*™ databases.

By default, when the edit box does not contain any text, and all databases are available, the lower section of the **Find** pane contains a list of solar system objects. When you type a name in the edit box at the top of the **Find** pane, the list updates to contain objects whose names match what you have typed. A number of controls are associated with each object shown in the match list.

75. Open the **Find** pane. Delete any entry that might be in the text box at the top of the **Find** pane.

76. Click the menu button (blue-colored button with an arrowhead icon) at the far left of an object whose name is not dimmed out in the list of solar system objects in the main part of the **Find** pane and select **Centre**. *Starry Night Enthusiast*™ finds and centers the object in the view. (You can press the **spacebar** to accomplish this centering immediately.)

77. In the list in the **Find** pane, click the checkbox to the immediate left of the object's name. *Starry Night Enthusiast*™ labels the object in the view.

78. Click the checkbox to the right of the object's name in the **Find** pane. *Starry Night Enthusiast*™ displays the object's orbit.

79. Click the **Info** symbol to the right of the object's name to display a summary of general information about the object below its entry in the match list.

80. Select the name of another object in the match list whose name is not dimmed out. Double-click the name of this object. *Starry Night Enthusiast*™ finds, labels, and centers the object in the view.

81. Repeat the previous step for an item in the match list whose name is dimmed out. A dialog box appears advising you that the selected object is not visible from the current viewing location on the time and date specified in the toolbar and asks if you would like to reset the time to when it can best be seen. If you click the **Best Time** button, *Starry Night Enthusiast*™ will alter the time in the toolbar and then find, label and center the selected object in the view.

82. Click the edit box in the **Find** pane and type **Polaris** followed by the **Enter** or **Return** key. *Starry Night Enthusiast*™ finds, labels, and centers Polaris, the North Star, in the view.

As you have seen, there are several ways to find and center an object in the view. Whichever method you use is a matter of choice and convenience.

K. Viewing Location

One of the unique features of *Starry Night Enthusiast*™ is its ability to show you views of the sky from other locations on Earth as well as from locations that are not available in real life. Often, views of the sky from other planets, stars, galaxies, or points in empty space will give new insight into an astronomical observation. In this section of the tutorial, you will learn how to get to some of these interesting locations.

83. Go **Home,** and **Stop** time flow.

84. Select **Options > Viewing Location...** (or use the keyboard shortcut **Ctrl-L**) to open the **Viewing Location** dialog box.

At the top of the **Viewing Location** dialog is the label **View From:** followed by two drop-down boxes containing the words **the surface of** and **Earth.** Below this is a series of tabs with the **List** tab highlighted. Below the tabs is a display pane showing a list of locations on the surface of the Earth.

85. Think of a city on the Earth that you have always wanted to visit. Type the name of the city on the keyboard. As you type, the list will skip to and highlight the name of the city that you typed. Alternatively, you can use the scrollbar to search the list and click on a city name of your choosing. With the name of the city highlighted, click the **Set Location** button. The dialog closes and the view adjusts to the view from the selected city.

86. Open the **Viewing Location** dialog again and click on the **Map** tab. Now the pane beneath the tabs shows a map of the Earth along with an indication of those regions of the Earth that are in daylight and those that are in night time. Click a spot on the map and then click **Set Location.** When the dialog closes and the view changes, the **Viewing Location** panel in the toolbar will tell you the location that you have selected.

87. Sometimes you want to specify a precise spot on the Earth (or another planet) in terms of Latitude and Longitude. Open the **Viewing Location** dialog once more and click on the **Latitude/Longitude** tab. The pane below the tabs contains edit boxes for entering the precise latitude and longitude to which you wish to go. When you click the **Set Location** button, *Starry Night Enthusiast*™ will take you to those precise coordinates.

88. Open the **Viewing Location** dialog and select **the centre of** in the first drop-box to the right of the **View from** label. Leave the selection in the second drop-box at **Earth.** Click the **Set Location** button. Your location changes to the centre of a transparent Earth. By adjusting the **Gaze,** you can explore the entire celestial sphere. Take some time to explore the celestial sphere from this unique location.

89. Open the **Viewing Location** dialog box and select **position hovering over** in the first drop-box to the right of the **View from** label. Click the **Set Location** button. The view looks down on your home location from 999 km above the surface of Earth. Change the **Time Flow Rate** to **3000×** in the toolbar. Observe the Earth rotating below you against a stationary background of stars.

90. Compare the view of Earth in the previous step with the view obtained by simply increasing the elevation of the viewing location. Click the **Home** button and **Stop time. Increase current elevation to about 1000 km.** Set the **Time Flow Rate** to **3000×.** Notice that in this case, the sky rotates over a stationary Earth. An elevated viewing location like this stays attached to its base point, so that it is as though you were sitting on the top of a flagpole that extends out into space. In the hover mode activated above, your viewing location remains stationary in space at a fixed distance from the surface and the Earth rotates beneath you.

91. Click the **Home** button and **Stop time.** Set the **Gaze** to North. Open the **Viewing Location** dialog box and select **position moving with** in the first drop-box to the right of the **View from** label. In the panel of options that appears, select **Above orbital plane** and then click the **Set Location** button. In this mode, if you set the **Time Flow Rate** to **30,000×** and **Run time forward,** you will see the Earth rotating and the stars drifting toward the right. In this view, you are at a location that is fixed relative to the Earth's position in its orbit around the Sun.

92. Open the **Viewing Location** dialog box and select **the surface of** in the first drop-box to the right of the **View from** label. Select **Mars** in the second drop-box. Click the **Set Location** button. Use what you have learned in previous sections of this tutorial to explore the sky as seen from the planet Mars.

93. Click the **Home** button in the toolbar to return to Earth.

L. Measuring Angles and Distances

A feature of *Starry Night Enthusiast*™ that is crucial to many of the observing projects is the **Angular Measurement Tool.** It is really two tools in one and it takes some practice to use this feature properly.

94. Click **Home.**

95. **Stop time.**

96. Change the **Time** to **12:00:00 AM.**

97. Locate a bright object (planet or star) in the view. Position the mouse cursor over this bright object. When the cursor changes its shape to that of the **Selection Tool,** click and hold the mouse button. If the cursor changes back into the **Hand Tool,** release the mouse button and try again.

98. While holding the mouse button down, drag the cursor to another object in the sky. As you move the cursor, a line appears from the original object to the current cursor position, along with a message indicating the angular separation between the two endpoints of this line.

99. As you drag the **Angular Measurement Tool** across the view, the cursor may occasionally snap to a nearby object. When it does so, two values are shown. The top, in red, is the angular separation between the two objects as seen from the current viewing location. The lower value, in blue, is the absolute distance between the two objects connected by the **Angular Measurement Tool.**

100. Select **Options > Viewing Location...** and choose **Calgary, Canada,** from the list of locations on the Earth. Then click the **Set Location** button to change your viewing location to this city.

101. Change the **Time** to **12:00:00 AM** and the **Date** to **June 1, 2007 AD.**

102. Change the **Gaze** direction to the South.

103. From the main menu, select **Labels > Planets-Moons.**

The view shows the sky looking south from Calgary, Canada, at midnight on June 1, 2007. The Moon, Jupiter, and Pluto are labeled.

104. Use the **Hand Tool** to adjust the view so that Jupiter and the Moon are visible in the view.

105. Use the **Angular Measurement Tool** to measure the angular separation between the Moon and Jupiter.

Question 5. What is the angular separation between Jupiter and the Moon on June 1, 2007?

Question 6. What is the actual absolute distance between the Moon and Jupiter on June 1, 2007?

M. Options and Presets

There are many options available through the *Starry Night Enthusiast*™ interface. There are options for changing the font, color, brightness, and size of various labels. You can display various astronomical guides and grids. You can change the way the view displays planets, stars, and moons. You can access these many options from the **Options** menu. An alternative way to access the options that are important to the observing projects is through the **Options** pane.

106. Click the **Home** button and **Stop time.**

107. Click the **Options** tab on the left border of the main view window to open the **Options** pane. Other ways to open the **Options** pane are: i) select **Options > Show View Options Panel** from the main menu, or ii) use the keyboard shortcut **Ctrl-J** (**Cmd-J** on a Macintosh system).

The **Viewing Options** pane is divided into seven sections. You can expand and collapse the section layers by clicking the icon to the left of the section heading. In each section, you will find a list of options. The first checkbox on the left in a list item indicates whether or not that option is in effect in the view. Next is the name of the object or option. If you move the mouse over the name of an option in the list, a button may appear that opens a dialog box for setting certain options pertinent to that item.

108. The items under the **Guides** section are various astronomical indicators and grids that *Starry Night Enthusiast*™ can overlay on the view. Click the checkbox to the left of the name of the indicator to turn it on (checked) or off (unchecked). To explore these guides, turn them on one at a time in the **Options** pane and then look around in the view to see the effect.

In the real world, factors such as the break of day, clouds, or the setting of an object below the horizon can interrupt an astronomical observation. Light pollution can degrade a real astronomical observation. *Starry Night Enthusiast*™ allows you to remove daylight, clouds, the horizon, and the effects of light pollution from the view.

109. Set the **Time** in the toolbar to noon, **12:00:00 PM.** The view will show a daylight sky. Click the checkbox to the left of the option labeled **Daylight** under the **Local View** section. Observe the result in the view. Then turn **Daylight** on again.

110. Click the checkbox to the left of the **Local Horizon** option under **Local View.** The horizon becomes transparent. Turn the **Local Horizon** on again. Point the cursor over the **Local Horizon** label and click the button that appears. The **Local Horizon Options** dialog box opens. In the dialog, click the **Select Clouds Panorama...** to open a list of cloud panoramas. Highlight the option in the lower left of the list labeled **None** and then click the **Select** button to return to the **Local Horizon Options** dialog. You may wish to change your horizon panorama at this point in time.

111. To see a simulation of the effect of light pollution on the night sky, the **Local Horizon** option must be checked. Change the **Time** to **12:00:00 AM.** Turn the **Gaze** to the North. Then click the **+** icon to the left of the **Distant Light Pollution** option under the **Local View** section of the **Options** pane. The layer expands to show two options. Click the checkbox next to the **Distant Light Pollution** option and then select **Large city in North.** Change the **Gaze** to **South** and check the **Smaller city in South** option. Switch the **Gaze** back and forth between South and North (*Hint:* the easiest way to do this is to use the S and N keys on the keyboard). When you are done, click the checkbox for **Distant Light Pollution** to turn this option off (unchecked).

Under the **Solar System** layer of the **Options** pane, you can set various attributes for the way solar system objects are displayed and labeled in the view.

112. Expand the **Solar System** layer of the **Options** pane.

Each item in the **Solar System** section is preceded by a checkbox that toggles the display of objects of this type on and off in the view. Clicking on the label of any of these items opens a dialog box in which you can change the settings for such things as how objects of this type are labeled in the view.

113. Click the **Planets-Moons...** label to open the **Planets-Moons Options** dialog box. Study the options in this dialog box so that you are familiar with them. Pay particular attention to the group of options in the **Other** category. We recommend that you leave the first option, **Enlarge Moon size at large FOVs,** checked. This option artificially enlarges the size of the Moon in the view when the field of view is wide. When you **Zoom In** to smaller fields of view, the Moon is displayed with its proper size relative to the field of view.

114. We recommend that you turn off the option to **Show solar lens flare when looking at Sun.** Similarly, select **Never** in the drop-box next to the **Sun halo** option. Though interesting, these effects can interfere with the accuracy of some of the observations you will make.

115. Experiment with the other options that you find in the **Options** pane and configure them to your satisfaction. Then, select **Options > Save Current Options as Default** from the menu bar. Now, each time you start *Starry Night Enthusiast*™, it will be configured with the options you have selected. You can always revert to the default options by selecting **Options > Presets > Default** from the main menu.

N. Preferences

Starry Night Enthusiast™ treats certain options as global preferences that influence the behavior of the interface. In this section of the tutorial, we will make recommendations on setting preferences that will make your use of *Starry Night Enthusiast*™ more efficient. You can always restore the global preferences that initially accompanied *Starry Night Enthusiast*™ by choosing **Preferences** from the **File** menu (Windows) or the **Starry Night Enthusiast** menu (Macintosh) and clicking the **Factory Defaults** button in the **Preferences** dialog box.

116. Select **File > Preferences** (select **Starry Night Enthusiast > Preferences** on a Macintosh system) from the main menu. This will open the **Preferences** dialog box.

117. Select **OpenGL** in the drop-box in the upper right corner of the **Preferences** dialog window. We recommend that you check the **Use OpenGL** if your system supports it. To speed transitions in the view, set the **Cross fade transition timing** slide control all the way to the left (0.00 seconds).

118. Select **Responsiveness** in the drop box in the upper right corner of the **Preferences** dialog window. Click the checkboxes next to the options labeled **Pan to found objects** and **Animate changes in location** so that these options are **off** (unchecked).

119. Select **Cursor Tracking (HUD)** in the drop box in the upper right corner of the **Preferences** dialog window. The **Show** panel is a list of all of the information that you can choose to display in the **HUD.** As you work through the observing projects there are times when you will need or want to change these options but for now we recommend checking only the following options: **Apparent magnitude, Constellation name, Distance from observer, Name,** and **Object Type.** You may also elect to show the HUD in the upper left corner of the view by checking the box below the **Show** panel.

120. Close the **Preferences** dialog box.

O. Files, Windows, and Favourites

You can save any view that you configure in *Starry Night Enthusiast*™ by selecting **Save** or **Save As…** from the **File** menu. You can use the standard operating system techniques to name the file and save it in an appropriate directory. You can then retrieve the view later when you select **Open…** from the **File** menu. When you open a file, or when you select **File > New…** from the main menu, a new instance of *Starry Night Enthusiast*™ launches. You can switch between each instance by selecting the name of the view you want to activate under the **Window** menu.

Favourites are *Starry Night Enthusiast*™ files or views that are saved to a special folder in the *Starry Night Enthusiast*™ program's directory structure. Favourite files are special in that they can be accessed easily from the **Favourites** menu or the **Favourites** pane. Moreover, when you select a Favourite by either of these methods, rather than opening a new *Starry Night Enthusiast*™ interface, the Favourite replaces the current view in the *Starry Night Enthusiast*™ interface from which the Favourite was selected.

While you have learned a great deal so far about configuring views in *Starry Night Enthusiast*™, you will be happy to know that many of the configurations that you will use in the observing projects are included as Favourites in the version of *Starry Night Enthusiast*™ that is bundled with the textbook *Universe*, 8th Edition, by Freedman and Kaufmann. **In the authors' experience, opening a view from the Favourites menu does not always show the initial view reliably. When this is the case with the files used in the projects in this book, we will explicitly instruct you to open the view from the Favourites pane.** Nevertheless, it is important that you know how to configure views in *Starry Night Enthusiast*™ in order to gain a better understanding of these pre-configured Favourites.

IMPORTANT

Any time that you exit *Starry Night Enthusiast*™ or use **File > Close** to terminate a view, a dialog box appears asking if you want to save the changes that you made to the view. Generally, and particularly with files designated as **Favourites,** you will want to click the **Don't Save** button. However, to prevent your Favourite files (particularly those that you require for the observing projects) from being corrupted by inadvertently clicking the **Save** button when this message appears, you or your computer administrator should ensure that all of the files with the .snf extension in the folder: **Program Files/Starry Night Enthusiast 5/Sky Data/Go/Observing Projects** are flagged as **read-only** files in your operating system.

The following sequence of instructions will introduce you to the convenience of using Favourites and also guide you in configuring and saving your own Favourites.

121. Select **Favourites > Guides > Atlas** from the main menu. The *Starry Night Enthusiast*™ file named **Atlas** loads into the view.

122. Select **File > New** from the main menu. A new instance of *Starry Night Enthusiast*™ with the name **Untitled1** in the title bar opens. Select the **Window** menu to see a list of both *Starry Night Enthusiast*™ instances that are currently open, namely the Atlas view and the view named Untitled1. Select **Window > Untitled1** from the main menu to make that view active. Change the **Gaze** in the **Untitled1** view so that you are looking **West** and then click the **Sunset** button in the toolbar.

123. With the view named **Untitled1** active on the screen, select **File > Close** from the menu. When the message appears, asking whether you want to save the changes you made to the view, select **Save**. In the **Save** dialog, select an appropriate folder into which you want to save this view, rename the view **Sunset 1** and click **Save**. The window closes and *Starry Night Enthusiast*™ returns you to the other remaining *Starry Night Enthusiast*™ window, the pre-configured **Atlas** view.

124. In the **Atlas** view, select **File > Open** and use the **Open** dialog to navigate to the **Sunset 1** file that you saved in the previous step and click the **Open** button. The view you saved as Sunset 1 opens in a new *Starry Night Enthusiast*™ window.

125. At this point, you can use standard techniques to resize the two windows so that both are visible on the screen simultaneously.

126. Select **Window > Sunset 1** from either *Starry Night Enthusiast*™ window. *Starry Night Enthusiast*™ makes the Sunset 1 view the active view.

127. In the Sunset 1 view, select **Guides > Precession** from the **Favourites** pane. Notice that *Starry Night Enthusiast*™ does not invite you to save any changes that you may have made to the Sunset 1 view and immediately loads the *Starry Night Enthusiast*™ **Precession** view. When you have finished watching the Precession view, select **File > Close** from the menu to close this instance of *Starry Night Enthusiast*™. IMPORTANT: When you close a view that was loaded from the Favourites menu, be sure to click the **Don't Save** button so that you do not overwrite these pre-configured files. As extra insurance against this eventuality, you or your computer administrator should flag all of the files with the **.snf** extension in the folder: **Program Files/Starry Night Enthusiast 5/Sky Data/Go/Observing Projects** as *read-only*.

128. When you close the **Precession** view, *Starry Night Enthusiast*™ returns you to the **Atlas** view. In the next steps, you will make a few changes to this view and save it as a Favourite named **My Atlas View.**

129. Open the **Options** pane and turn off the display of all of the items under the **Guides** section. Select **Labels > Hide All Labels** from the menu. To save this version of the **Atlas** view as a Favourite, select **Favourites > Add Favourite** from the main menu. The Favourites pane opens to the left of the view and a new entry appears in the list. Name this new favourite **My Atlas View.**

130. Notice that the new Favourite that you named **My Atlas View** can be loaded from the **Favourites** pane but does not appear under the **Favourites** menu.

131. Close the *Starry Night Enthusiast*™ application, selecting the **Don't Save** button if asked whether you want to save any changes you may have made to the currently active view.

132. Now, re-launch *Starry Night Enthusiast*™ and look under the **Favourites** menu. You should find the new Favourite named **My Atlas View** in the list of Favourites under this menu, now that you have restarted *Starry Night Enthusiast*™.

P. Backtracking

There may be times when you are working in a view and issue a command and then want to reverse that action to return the view to the previous state. To do this, you can select **Edit > Undo** from the menu. Alternatively, you can use the keyboard shortcut **Ctrl-Z** (**Cmd-Z** on a Macintosh). By default, you can reverse up to 50 previously issued commands by using the **Undo** feature.

At other times, you might want to revert to the view with its initial opening conditions. To do this, you can select **File > Revert** from the menu. The following sequence of instructions demonstrates these techniques for you.

133. Select the view that you named **My Atlas View** from the **Favourites** pane.

134. Open the **Options** pane and turn on the **Celestial Grid** option under the **Guides** layer. The view now includes a grid overlaid on the sky.

135. Select **Edit > Undo Guide Setting** to reverse the previous step.

136. Issue several other commands to *Starry Night Enthusiast*™ to make changes to the view named **My Atlas View.** To return to the initial state of this view, select **File > Revert** from the menu. The view returns to its initial conditions.

137. At this point, you may want to delete the Favourite you created named **My Atlas View.** To do so, open the **Favourites** pane, highlight the title of the view and click the right mouse button (hold down the **Ctrl** key while clicking the mouse button on a Macintosh). Select **Delete File** from the menu that appears and then click the **Delete** button in the message box that appears.

138. Exit the *Starry Night Enthusiast*™ program at this point, making sure to select the **Don't Save** button when asked if you want to save the changes you made to the **My Atlas View** file.

Q. Put It All Together

The goal of this tutorial has been to introduce you to the features of *Starry Night Enthusiast*™ and make you feel comfortable with the interface. In the next sequence of instructions, you will get a chance to see the extent of the *Starry Night Enthusiast*™ universe.

139. Launch *Starry Night Enthusiast*™.

140. **Stop time.**

141. Change the **Viewing Location** to a **position hovering over** the **North Pole** of Earth. [The easiest way to do this is to click on the top center of the map that appears in the **Viewing Location** dialog box after you select the hovering option.]

The view shows the Earth from a location about 1000 kilometers above its surface.

142. **Increase current elevation** approximately 10 times to about **10,000 km.**

143. Set the **Time Flow Rate** to 3000× and watch the Earth rotate beneath you.

144. **Increase current elevation** by another factor of about 10 times to about **0.001 AU from Earth.**

145. **Increase current elevation** by another factor of ten to **0.010 AU.**

146. Select **Labels > Planets-Moons** from the main menu and then change the **Time Flow Rate** to 30,000× and watch the Moon revolve around the rotating Earth.

147. **Increase current elevation** by another factor of ten to **0.100 AU.**

148. Change the **Time Flow Rate** to **1 day** and click the **Run time forward** button in the toolbar.

149. **Increase current elevation** by another factor of ten to about **1 AU.**

The Sun, Mercury, Venus, and Mars occasionally enter the view. Because the view is centered and locked on the Earth, these objects appear to move around the Earth in a complicated fashion. Before proceeding further out into space, change the view so that it is centered on the Sun.

150. **Stop time.**

151. Select **Options > Viewing Location...** from the main menu and change the **View from:** setting in the dialog window to **Stationary location.** Then click the **Set Location** button.

152. Open the object contextual menu for the **Sun** (Windows users click the right mouse button over the Sun. Macintosh users hold down the **Ctrl** key and click on the image of the Sun.) Select **Centre** from the drop down menu.

153. Open the **Find** pane and click the checkboxes to the right of the names of the planets in the list. This will display the orbits of the planets in the view.

154. **Run time forward** to watch the inner planets of the solar system orbit the Sun.

155. Click the menu button to the left of the listing for the **Earth** in the **Find** pane and select **Centre** from the menu to center the gaze back onto the Earth.

156. **Increase current elevation** by another factor of 10 to about **10 AU from Earth.**

157. **Increase current elevation** by another factor of 10 to about **100 AU from Earth.**

158. **Increase current elevation** by another factor of 10 to about **1000 AU from Earth.**

159. **Increase current elevation** by another factor of 10. Notice that before reaching a distance of 10,000 AU from the Earth, the display in the **Viewing Location** pane in the toolbar switches to distances in light-years. Adjust the elevation so that you are about **0.1 ly from Earth.**

160. Watch the stars to the right of the Sun as you gradually **Increase current elevation** by another factor of 10 to about **1 ly from Earth.** Note that a bright star appears to be moving toward the Sun. Use the **HUD** to identify this star.

Question 7. What is the name of the star that appears to shift its position markedly as you increase elevation?

161. Use the **Hand Tool** to measure the spacing between this star and the Sun.

Question 8. What is the distance between this star and the Sun in light-years?

The motion of this star is a result of an effect called parallax, in which relatively close objects appear to move against the more distant background as you change your viewing location.

Question 9. By what total factor have you increased your distance from the initial location, where you were hovering 999 kilometers above the Earth? [*Hint:* Count the number of steps so far in which you increased your elevation by a factor of 10 and raise the number 10 to the power of this number.]

162. **Increase current elevation** by another factor of 10, to about **10 ly from Earth.**

Other stars in the view start to show the parallax effect.

163. **Increase current elevation** by another factor of 10, to about **100 ly from Earth.**

164. **Increase current elevation** by another factor of 10, to about **1000 ly from Earth.**

The Sun is labeled and surrounded by a tight cluster of stars. These are the stars in the Milky Way Galaxy that are relatively close to the Sun. They are visible with the naked eye from the Earth in the night sky and form the patterns of the constellations. You will also see a partial image of our Galaxy, the Milky Way.

165. **Increase current elevation** by another factor of 10 to about **10,000 ly from Earth.**

Other points of light become visible in the left side of the view. Each point represents one of the 28,000 galaxies near the Milky Way that are included in the database of this version of *Starry Night Enthusiast*™.

166. **Increase current elevation** by another factor of 10 to about **0.100 Mly from Earth.** At this distance, 0.1 million light-years from the Sun, the entire Milky Way is visible in the view.

167. **Increase current elevation** by another factor of 10 to about **1 Mly from Earth.**

168. **Increase current elevation** by another factor of 10 to about **10 Mly from Earth.**

169. **Increase current elevation** by another factor of 10 to about **100 Mly from Earth.**

170. Finally, **Increase current elevation** by another factor of 10 to about **1000 Mly from Earth.** This is equivalent to a distance of 1 billion light-years from the Earth.

171. Use the **Location Scroller** to change the point of view.

At this distance from the Sun, you see the extent of the *Starry Night Enthusiast*™ database of 28,000 of the closest galaxies to the Milky Way. In reality, the observable universe, which extends out from the Earth by another factor of more than 10 to about 12 billion light-years, is filled with billions of galaxies and you should keep this in mind as you look around the view.

172. Click and hold the **Decrease current elevation** button in the toolbar to move rapidly back down to the surface of the Earth. You may need to release the button occasionally to allow the view to catch up with this rapid flight back to Earth.

R. Observing a Famous Visitor

Use the next sequence of instructions to find a famous celestial visitor to Earth.

173. Set the **Viewing Location** to Alice Springs, Australia.

174. **Stop time** and set the **Time** and **Date** to **12:00:00 AM** on **April 5, 1986.**

175. Set the **Gaze** to the **Southeast** (azimuth 135°) and use the **HUD** to identify the object over the southeast horizon.

Question 10. What is the name of the interesting object in the view?

176. Select **File > Save as...** to save this view with the title **Comet** in a suitable folder.

177. Open the object contextual menu for this comet and select **Centre** from the list of options.

178. **Zoom In** to a field of about **25° × 17°.**

179. Change the **Time Flow Rate** to 3000× and observe whether the comet moves relative to the background stars.

180. Select **Edit > Undo Time Step (Ctrl-Z)** to return to the start of the animation and select **Edit > Redo Time Step (Ctrl-Shift-Z)** to see it again.

Question 11. Does the comet move relative to the stars or does it remain stationary?

181. To verify your answer to the previous question, select **File > Revert** to return to the initial conditions. Change the **Time Flow Rate** to **1 sidereal day** and then **Step time forward.**

S. Fly Solo!

On July 20, 1969 AD, at 3:17:40 PM Central Daylight Time, Neil Armstrong and Buzz Aldrin landed their lunar module, *Eagle,* at Tranquility Base on the Moon. Try to configure *Starry Night Enthusiast*™ to show a view of the Moon as it appeared at that time from Mission Control in Houston, Texas (Central Time Zone).

> **Question 12.** a) Where was the Moon in the sky as seen from Houston on the afternoon of July 20, 1969? (Direction? Azimuth? Elevation?) b) At what phase was the Moon at this time?

As a final test of your expertise with *Starry Night Enthusiast*™, attempt to simulate the view of the Earth that the astronauts would have seen, had they looked through the window of their vehicle toward the Earth after landing at 20:17:40 UT (Universal Time) on July 20, 1969, and again as Neil Armstrong would have seen the Earth as he stepped onto the lunar surface at 02:56:15 UT on July 21, 1969. [*Tip:* The **Viewing Location** dialog box contains a listing of the locations of each of the Apollo landing sites.]

> **Question 13.** Which hemisphere of Earth was directed toward the Moon at the time that Neil Armstrong and Buzz Aldrin landed on the Moon?

> **Question 14.** Which continent of Earth would have been visible to Neil Armstrong at the historic moment that he stepped onto the lunar surface and announced: "That's one small step for [a] man...one giant leap for mankind."

T. Conclusion

While this tutorial has introduced you to some of the many features of *Starry Night Enthusiast*™, it is by no means exhaustive. You should certainly read the User's Guide that is available under the **Help** menu to acquaint yourself with *all* of the features of *Starry Night Enthusiast*™ software that were not discussed in this tutorial. Explore the **Starry Night features** and **Starry Night basics** options in the **Sky Guide** side pane.

Stars and Constellations 2

This project will familiarize you with several of the major constellations and bright stars that can be seen from mid-northern latitudes on the Earth. Once you know these patterns and can find them in the night sky, you can use them to locate other stars and constellations. With practice, you can become expert at finding your way around the sky.

Historically, a constellation is a group of stars that outlines, or at least represents in some fashion, a familiar object or pattern. For example, the bright stars in the constellation of Orion roughly outline the body of Orion, the hunter, while the long line of stars making up the constellation Draco resemble the curved shape of a dragon. Many of these names have their origins in the myths and stories of antiquity.

In modern astronomy, constellations are defined differently. Each constellation is an area of the sky with precisely defined boundaries, rather than a group of stars. Together, the 88 officially designated constellations cover the entire celestial sphere. In this way, the constellations provide us with an atlas of the celestial sphere. Thus, when we say that the star Regulus is in the constellation of Leo, this tells us the location of Regulus on the sky in the same way that saying that Grenoble is in France tells us where Grenoble is on the Earth. However, most people—including most astronomers—picture a constellation more often in terms of the historical star pattern than as a set of invisible boundary lines dividing the sky into sections.

A. Constellations

In this section you will use the *Starry Night Enthusiast*™ program to understand the difference between the historical and modern definitions of constellations.

1. Launch *Starry Night Enthusiast*™.

2. Select **File > Preferences (Starry Night Enthusiast > Preferences** on a Macintosh system) from the menu. In the **Preferences** dialog, choose **Cursor Tracking (HUD)** in the drop box. In the **Show** list, choose only the options **Constellation name, Name,** and **Object type.**

3. **Stop Time** and change the **Time** in the toolbar to **12:00:00 AM Standard Time,** at your home location.

4. Select **View > Constellations > Astronomical** and **View > Constellations > Labels** from the menu.

The view shows the familiar historical definition of the constellations. Stars within a constellation are grouped to form a recognizable pattern.

5. Use the **Hand Tool** or **scrollbars** (select **View > Show Scrollbars**) to scroll across the sky to survey the various constellations visible from your home location.

6. Open the **Options** pane and, under **Local View**, click the checkbox to the left of **Local Horizon** to remove the horizon from the view, and use the **Hand Tool** or **scrollbars** to look at all of the constellations on the celestial sphere.

7. Open the **Options** pane, expand the **Constellations** layer, and click the checkbox to the left of the label **Boundaries** to select it and click the checkbox next to the label **Stick Figures** to turn this option off.

Now the sky is divided by boundaries like countries on a map. Each "country" shows the modern astronomical definition of a constellation. A constellation contains all of the stars within its boundaries regardless of whether these stars are also included in the more familiar historical pattern of the constellation.

Question 1. Name three of the smallest constellations on the celestial sphere.

Question 2. Name three of the largest constellations on the celestial sphere.

Question 3. What is the approximate length in degrees of the largest constellation on the celestial sphere? (Hint: The maximum field of view is 100° wide.)

The brighter stars within a constellation are generally ranked according to their apparent brightness with Greek letter prefixes. For example, the brightest star in the constellation Leo, the Lion, is called Alpha Leonis; the next brightest star is called Beta Leonis, and so on. Often, these prominent stars will also have common names. Alpha Leonis, for instance, is also known as Regulus.

A third display option is available showing an artist's impression of imaginary images representing the constellations.

8. In the **Options** pane under the **Constellations** layer, click the checkbox next to the label **Illustrations** and click off the **Boundaries** option.

9. Use the **Hand Tool** or **scrollbars** (select **View > Show Scrollbars**) to scroll across your local sky to view these images.

These striking representations may help you in remembering the form of the constellations when you come to explore the real sky with none of these guides available.

The table below shows the optimum times for looking for major constellations or well-known collections of stars called asterisms from mid-latitude northern hemisphere sites.

Constellation or Asterism	Time of Best Visibility
Big Dipper	February to August
Cassiopeia	August to March
Auriga	November to April
Bootes	April to August
Virgo	April to July
Leo	March to June
Orion	December to April

B. Asterisms

Some prominent patterns of stars that have common historical names are not constellations. These are called **asterisms**. Some asterisms are part of a constellation. An example is the Big Dipper asterism, which forms a part of the constellation Ursa Major. Other asterisms extend over two or more constellations. For example, the summer triangle asterism consists of three stars, each from a different constellation. The next steps will help you to understand the difference between asterisms and constellations.

10. Select **Favourites > Observing Projects > Stars and Constellations > Big Dipper.**

The view shows the night sky at midnight on June 4, 2006, from Calgary, Canada, at latitude 51°, looking west at about 45° above the horizon. Constellation patterns, labels, and modern boundaries are visible. Notice Ursa Major, commonly known as the Big Bear, to the right of center in the view.

11. Select **View > Constellations > Asterisms** to turn this option on.

The view now shows the asterisms that are found in this part of the sky. You will see that the Big Dipper asterism, so named because its shape resembles that of a scoop or dipper, is within the boundaries of Ursa Major but forms only a part of the constellation's more extensive pattern. Notice also the large asterism named the Diamond of Virgo. Each of the four stars comprising this asterism is within a different constellation.

12. Toggle back and forth between the view of the asterisms and the view of the traditional constellations and their modern boundaries by selecting **File > Revert** and the **Edit > Undo Revert** to answer the following questions.

Question 4. Which constellations contribute stars to the Diamond of Virgo asterism?

Question 5. What are the names of the four stars that make up the Diamond of Virgo asterism?

Question 6. Which constellation is almost completely encompassed by the Diamond of Virgo asterism but does not contribute a star to the asterism's pattern?

As the Earth rotates, the orientation of constellations and asterisms varies with respect to compass directions on Earth. Their visibility will depend on your location on the Earth and the season. Nevertheless, their easily recognizable patterns make asterisms and constellations invaluable guides for finding your way in the night sky.

C. Finding the Big Dipper

A very useful guide to the sky for northern hemisphere observers is the above-mentioned Big Dipper asterism within the constellation Ursa Major, the Great Bear. This asterism is in the shape of a spoon or scoop that might be used to serve soup or scoop water from a barrel. It is made up of seven bright stars. Three of them in the tail of the bear make up the handle while the other four make up the scoop and are in the body of the bear. If you live between latitudes of about 40° N and 60° N and want to find the Big Dipper asterism in the actual sky, face approximately north and look at a point about halfway up from the horizon. Imagine a large circle centered on this point, starting near the northern horizon and curving up to the left to a point near the zenith, and then curving down to the right, back to the starting point. The easily recognized pattern of the Big Dipper should be somewhere along or near this circle. If you live south of 40° N latitude, the Big Dipper may be below the horizon during certain times of the year or at certain times of the night. Consult the previous table for the best times to view the Big Dipper. If you live north of 60° N latitude, the imaginary circle described above will include points to the south of the zenith.

The Big Dipper can have any orientation, including upside down or hanging downward from its handle, depending on the time of night and the time of year. The important thing is to be able to recognize its pattern.

13. Select **Favourites > Observing Projects > Stars and Constellations > Constellations.**

The view is of the northern sky from New York City at 10:00 PM EDT on September 1, 2006. Using the technique described above, you should quickly find the Big Dipper near to the left side of the view. The seven bright stars that form the shape of a dipper comprise the Big Dipper. At this time, the handle extends away from the bowl toward the upper left of the screen.

14. To confirm that you have correctly identified the Big Dipper in the view, select **View > Constellations > Asterisms** and **Labels > Constellations** from the menu.

Question 7. In addition to the Big Dipper, which other asterisms are visible in the view?

15. Once you have identified the Big Dipper in the view, select **Labels > Constellations** to turn the labels off again.

D. Describing Directions from the Big Dipper

The Big Dipper is an important asterism to locate because its member stars can be used as pointers to many other stars and constellations. For example, two of this asterism's stars provide a pointer to the North Star. The ability to find the North Star is very useful because it always lies directly above the north point of the horizon, at least at this time in history—a valuable guide when orienting oneself on the Earth.

In order to use the Big Dipper as a guide to other constellations, asterisms, and stars, we first need to define the directions "upward" and "downward" relative to the Big Dipper. Imagine that the dipper shape is oriented exactly horizontal with the bottom of the bowl resting on a flat table and the handle extending toward your left. In this orientation, the bowl opens upward, and it is in this sense that we shall use the term "upward." Conversely, downward is the direction in which water would drip if the bowl of the dipper leaked. Remember, the bowl of the Big Dipper always opens "upward" toward the north celestial pole no matter which orientation the Big Dipper has in the sky.

16. Change the **Time** in the toolbar to **12:00:00 AM EDT** in order to see the orientation of the Big Dipper as described in the previous paragraph.

17. Press the **K** key to turn off the outline of the asterisms in the sky.

18. Click on **View > Solar System > Satellites** to remove the confusion of Earth-orbiting satellites from the view.

19. Change the **Time Flow Rate** in the toolbar to **3000×** and watch the changing orientation of the Big Dipper in the night sky as time progresses at 3000 times its normal rate.

20. **Stop Time** at approximately **5:00 AM EDT** on **September 2** as daylight begins to interfere with the view.

With time flowing forward at this advanced rate, you see the change in the orientation of the Big Dipper with respect to the horizon as the sky rotates around the North Star due to Earth's rotation. By 5:00 AM, the Big Dipper is standing on its handle on the right side of the view. In this orientation, "upward" from the bowl, as defined above, is toward the upper left in the sky and "downward" is toward the lower right.

It may seem odd to use the word "upward" in this sense; however, it allows us to use the Big Dipper as a guide to the sky without worrying about how the Big Dipper is oriented with respect to the horizon.

E. Finding the North Star

To find Polaris, the North Star, using the Big Dipper as a guide, first locate the handle of the Big Dipper and then find the two stars that are on the end of the bowl farthest from the handle. These two stars are often called the "pointer stars" because they point to the North Star.

21. Select **File > Revert**.

22. Position the **Hand Tool** over the bottom pointer star (the star at the bottom of the bowl farthest from the handle). Use the **Angular Measurement** facility of the **Hand Tool** to draw a line directly through the top pointer star (the star at the top of the bowl) and extend the line a distance approximately equal to seven times the distance between the two pointer stars, or about 35°.

23. Release the mouse button to deactivate the **Angular Measurement Tool** and position the **Hand Tool** over the nearest bright star to display its name, Polaris, the North Star.

The pointing is not perfect, but it is fairly close. While the North Star is not a really bright star, it is the brightest star in that little patch of the sky. At this time in history, this star is close to the north celestial pole, the point in the sky directly above the north end of the spin axis of Earth.

Question 8. What are the names of the two "pointer" stars in the Big Dipper?

Question 9. What is the actual angular distance between the bottom star of the pointers and the Pole Star?

Interesting Objects in Ursa Major

Before going on to find further constellations, it is interesting to look more closely at two objects in Ursa Major. The first is the second star from the end of the handle of the Big Dipper, named Mizar.

24. Select **File > Revert**.

25. Move the cursor over the second star of the handle of the Big Dipper, right-click (Macintosh users **Ctrl-click**) and select **Centre** to move this star to the center of the view. You will note that this star has a companion. With good vision, this pair of stars is easily separated with the unaided eye in the real sky.

26. **Zoom In** to a field of view of about 6° × 4°. The individual stars of this binary pair are clearly separated and a faint companion has become apparent, making a triangle of stars.

Question 10. What is the name of Mizar's bright companion?

Question 11. What is the separation of the two stars in this binary pair of stars?

27. **Zoom In** to a field of view about **40'** wide. At this magnification, you can see that Mizar has a companion star. Analysis of the spectrum of the light from these stars reveals that each of these stars is itself a binary star. Thus, the second "star" in the handle of the Big Dipper has been shown to be a small group of six stars! In fact, this is not uncommon in our part of the universe, with binary stars and small groups making up a large proportion of our neighborhood stars.

28. **Zoom In** to a field of view of about **1'** wide and measure the separation between Mizar and its companion.

Question 12. What is the name of Mizar's close companion star?

Question 13. What is the separation between the two stars in this binary pair of stars?

The second interesting region of this constellation contains two galaxies, M81 and M82. This M designation was assigned to diffuse objects in the sky by Messier to help him to avoid confusing them with comets for which he was searching.

29. Select **File > Revert** to return to the wide field of view of this constellation.

30. Open the **Options** pane and, under the **Deep Space** layer, click in the boxes on the left and right of the **Messier Objects** to display these galaxies and their labels. Bode's Galaxy, M81, and the Cigar Galaxy, M82, can be seen to the north and esst of the pointer stars at this time. Click off the Chandra Image to remove the X-ray image from M82.

31. **Centre** and **Zoom In** on one or other of these galaxies to a field of view of about 1 degree to see these very different galaxies. M81 is a beautiful spiral galaxy, while M82 shows evidence of violent activity and star formation fairly recently in galactic time. Both of these galaxies are faintly visible in binoculars in the real sky.

Question 14. What is the separation between these two galaxies in our sky?

F. Finding the Little Dipper

Polaris is the end star in the handle of the Little Dipper, an asterism in the constellation Ursa Minor. Starting from Polaris, you should be able to discern a line of four stars curving in an arc toward the handle of the Big Dipper. The first three of these stars are fainter than Polaris, and the fourth is about equal in brightness to Polaris.

32. Select **File > Revert** to return to the wide field of view.

33. Search for and identify the **Little Dipper**.

34. Select **View > Constellations > Asterisms** from the menu and then select **Labels > Constellations** to verify that you have found the Little Dipper. Then press the **K** key on the keyboard to turn the illustrations and labels off once more.

At the end of the Little Dipper nearest to the handle of the Big Dipper are two stars, Kochab and Pherkad. These two stars, which are brighter than all of the other stars in the Little Dipper except for Polaris, are often referred to as the Guardians. As the sky turns, the Guardians always remain between the Big Dipper and Polaris and "guard" Polaris from the Great Bear.

35. Use the **HUD** feature of the **Hand Tool** to identify the two Guardian stars, Kochab and Pherkad.

G. Finding Cassiopeia, or the "W"

36. Select **File > Revert**.

At this time, with the Big Dipper to the west of north in the view, the constellation Cassiopeia is found by imagining a straight line from the handle of the Big Dipper to the North Star extended by almost an equal distance past the North Star. The brightest stars in the constellation Cassiopeia (the Queen of Ethiopia) form an asterism shaped like a W. The end of this line should place you within this group of stars. In the orientation on your screen, the "W" is tipped up on one end.

37. Search for the constellation of **Cassiopeia**.

38. Select **View > Constellations > Asterisms** and then **Labels > Constellations** from the menu to verify that you have found the "W" asterism. Then turn these options off again by pressing the K key on the keyboard.

> **Question 15.** What are the names of the stars that make up the "W" asterism?

Interesting Objects in Cassiopeia

This constellation contains two open clusters of young stars, M103 and M52, the Scorpion.

39. Select **File > Revert**.

40. Open the **Options** pane and, under **Deep Space**, click on the box to the right of **Messier Objects** to label these objects. You can see that M103 is next to Ruchbah, one of the stars of the "W" while M52 is on the extension of the line between the right-hand two stars of the "W."

41. **Centre** on each of these objects in turn and **Zoom In** to examine these clusters. Note particularly the blue and red stars contained in M103.

> **Question 16.** Which of the two open clusters of young stars is the "richest" in having the higher number of stars?

H. Finding the Star Capella and the Constellation Auriga

Capella is the brightest star in the constellation Auriga, the Charioteer. To find this star, first locate the two stars on the top of the bowl of the Big Dipper. One of these two stars, Megrez, marks the point where the handle joins the bowl, and the other is the upper of the two pointer stars (Dubhe).

42. Select **File > Revert** to return to the wide field of view.

43. Beginning at the star Megrez, use the **Angular Measurement** feature of the **Hand Tool** to draw a line through Dubhe and extend it across the sky (toward the right on your screen) a distance of about six times the spacing between Megrez and Dubhe (about 60°). The *really* bright star near the end of this line is Capella. Use the **HUD** feature of the **Hand Tool** to verify your identification of this star.

Capella is actually slightly above the imaginary line you have drawn in the previous step, but it is the only really bright star that stands out in that part of the sky. When you look at Capella in the actual sky you may notice that it has a yellowish color. This is because Capella is roughly the same temperature as the Sun. However, while our Sun is a dwarf star, Capella is a supergiant star with a radius approximately 14 times larger than that of the Sun and is therefore intrinsically much brighter than the Sun.

The brightest stars of the constellation Auriga consist of five stars, including Capella, in the form of a stretched pentagon.

44. Position the **Hand Tool** over Capella and click the right mouse button (Macintosh users hold down the **Ctrl** key and click the mouse button over the star) to open the object contextual menu for this star and select **Centre**.

45. Change the **Time Flow Rate** to 3000× and watch as time runs forward to about **1:00:00 AM EDT** on **September 2, 2006.**

46. Select **View > Constellations > Astronomical** from the menu. Then select **Labels > Constellations.**

I. Finding the Star Arcturus and the Constellation Bootes

47. Select **File > Revert.**

48. Adjust the view with the **Hand Tool** or **scrollbars** so that the gaze direction is between **W** and **NW**, with the horizon low on the screen.

In this view, the Big Dipper is visible to the right in the view. To find the star Arcturus, start by looking at the handle of the Big Dipper. Notice that the handle is curved and forms an arc in the sky. Imagine continuing this arc through the sky in a direction away from the bowl of the Big Dipper for a distance roughly equal to the total length of the Big Dipper asterism. You should discover that this arc passes more-or-less through a bright star just below and to the left of the center of the view. This star is Arcturus. A good way to remember how to find this star is to "Follow the arc to Arcturus."

Arcturus is the brightest star in the constellation Bootes, the hunter. (The two "o's" in Bootes are pronounced separately: Bo-otes. Writing the name of the constellation as Boötes indicates this. The "e" is also pronounced.) If you look up and right from Arcturus in the view, you should see a pentagon of five stars. Arcturus plus the other five form a kite or ice-cream-cone shape in the sky.

49. Select **View > Constellations > Astronomical** and then **Labels > Constellations** from the menu to turn on the stick figures and labels of the constellations in the view to check that you have found Bootes correctly.

Question 17. Which stars (apart from Arcturus) make up the "kite" or "ice-cream-cone" shape of Bootes?

In the real sky, Arcturus has a yellowish or even slightly orange tint because it is slightly cooler than the Sun. It is also a giant star—much smaller than the supergiant Capella, but still much larger than our Sun.

With the constellation lines and labels on in the view, you can see another constellation, Canes Venatici, below the handle of the Big Dipper. Canes Venatici (Latin for "dogs of the hunter") consists of two relatively faint stars joined by a single straight line.

Question 18. What are the names of the two stars in Canes Venatici?

In mythology, Bootes is hunting the Great Bear. As the sky turns counterclockwise around the North Star, Bootes follows the Great Bear around the sky. The two stars in Canes Venatici are the two hunting dogs belonging to Bootes, nipping at the heels of the Great Bear as it circles the North Star, trying to get away from Bootes.

50. Select **View > Constellations > Illustrations** to see an illustration of these mythological figures. Turn this feature off again after you have examined these mythical and ancient figures.

J. Finding the Star Spica and the Constellation Virgo

51. Press the **K** key on the keyboard to turn off the stick figures, illustrations, and labels of the constellations.
52. Change the **Time** in the toolbar to **8:30:00 PM EDT** on **September 1, 2006**.
53. Press the **W** key on the keyboard to set the gaze direction due west.

In this view of the twilight sky, locate the arc of the Big Dipper in the upper right of the view (the complete Big Dipper asterism may not be visible in this view but the arc of its handle should be obvious). To find Spica, start at the handle of the Big Dipper and "follow the arc to Arcturus!" Continue this arc for roughly the same distance again past Arcturus to "speed on to Spica." Spica is the relatively bright star near the horizon, between the south and southwest compass points.

54. Use the **HUD** to identify Spica.

Spica is the brightest star in the constellation Virgo, the Virgin. Spica is much hotter than our Sun and when viewed in the actual sky, it can be seen to have a slightly bluish tint.

55. Select **View > Constellations > Astronomical** from the menu and then **Labels > Constellations** to see the pattern of stars in the constellation Virgo. Then turn these features off again by pressing the **K** key on the keyboard.

K. Finding the Star Regulus and the Constellation Leo

In order to find Regulus in the constellation Leo, the Lion, we need to move the date to December and change the time and viewing direction.

56. Change the **Date** in the toolbar to **December 1, 2005 AD**.
57. Turn **Daylight Saving Time** off (click the Sun icon in the **Time** display in the toolbar).
58. Change the **Time** in the toolbar to **1:00:00 AM EST**.
59. Use the **Hand Tool** or **scrollbars** to change the gaze to midway between **E** and **NE**, with the horizon near the bottom of the screen.

In this view, locate the Big Dipper, standing on its handle on the left side of the screen. To find the star Regulus, locate the two stars in the Big Dipper on the opposite side of the bowl from the pointer stars (i.e., the star where the bowl of the dipper meets the handle, Megrez, and the star Phecda to the right of Megrez in this view).

60. Starting at Megrez, use the angular measurement feature of the **Hand Tool** to draw a line in the direction of Phecda and extend it a distance about 10 times the distance between Megrez and Phecda (about 50°). The bright star at the right hand end of a small grouping of stars is Regulus. Use the constellation display and **HUD** features to verify your identification of the constellation Leo and its most prominent star, Regulus.

Regulus, like Spica, is hotter than our Sun and has a bluish tint when viewed in the real sky. Regulus is the brightest star in the constellation Leo. The most easily recognized part of this constellation is a line of stars running toward the upper left from Regulus forming the shape of a sickle or backward question mark.

61. To see the constellation Leo containing the Sickle asterism, select **View > Constellations > Asterisms** from the menu and then select **Labels > Constellations.**

The Sickle asterism represents the head of the lion in the constellation Leo. The hind end of the lion is formed by a triangle of stars below and to the left of the Sickle.

62. To see the rest of the constellation Leo, select **View > Constellations > Astronomical** from the menu.
63. Select **View > Constellations > Illustrations** to see an illustration of the mythical lion, Leo.

L. Conclusion and Suggested Extensions

This project has shown you how to find your way around the sky by using the Big Dipper as a guide. You should use the techniques discussed in this project to locate these stars and constellations in the real sky. Then, using a star atlas or printouts from your *Starry Night Enthusiast*™ software and maybe a pair of binoculars, you can explore other constellations to extend your knowledge of the night sky.

Astronomical Coordinate Systems 3

The concept of the **celestial sphere** is a simple but useful model of the universe. In this model, the sky is a vast spherical shell of infinite radius, centered on the Earth. All of the objects that appear in the sky are assumed to be so far away that they are equidistant from the Earth on the celestial sphere, rendering the sky as a two-dimensional spherical surface.

The spherical geometry of astronomical coordinate systems is less familiar than the Cartesian coordinate systems used for flat surfaces, but the basic concepts are equivalent. The position of an object on a two-dimensional surface, whether flat or spherical, requires only two coordinates. The first coordinate is the perpendicular distance to the object from a chosen reference plane that intersects the surface. The second coordinate is the distance to the object from a chosen reference point along this intersection of the reference plane with the surface.

> The celestial sphere is discussed in Section 2–4 of Freedman, Geller, and Kaufmann, *Universe*, 9th Ed.

One difference between the coordinate system of a flat surface and that of a spherical surface is the dimensions used in these coordinate axes. In the coordinate system of a flat surface, these dimensions are linear and are measured in units of length. In a spherical coordinate system, these dimensions are angular and measured in units of angle.

> Celestial coordinates are discussed in Box 2–1 of Freedman, Geller, and Kaufmann, *Universe*, 9th Ed.

The reference plane and reference point of a coordinate system are chosen to make it convenient for a particular purpose. In this project, you will explore methods for specifying the position of objects in the sky based upon several specific reference frames using the model of the celestial sphere.

A. Spherical Geometry

Figure 1 shows the sky modeled as a spherical shell, O being the center of the sphere, as labeled in the figure. Any plane containing O will intersect the sphere on a circle defined as a **great circle.** One specific great circle is chosen as a reference plane for a spherical coordinate system and is known as the **equator** because it divides the sphere into two equal hemispheres.

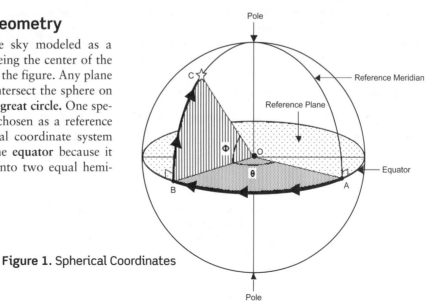

Figure 1. Spherical Coordinates

A line passing through the center of the sphere perpendicular to the reference plane intersects the sphere at two points called **poles.** Each pole is at the center of its respective hemisphere. Any plane perpendicular to the equator passing through the center of the sphere will include both poles and intersect the sphere in specific great circles known as **meridians.**

We can select a reference point on the equator, point **A** in the figure, such that the meridian through **A** becomes the reference meridian. Any position of an object on the sphere, such as the star at point **C,** can now be specified by two measurements. The first of these is the angular distance along the equator from the reference point **A** to the meridian passing through the object, point **B** in the figure. The second is the angular distance of the object from the equator, as measured along the object's meridian, from point **B** to point **C** in the figure.

Note that angular arcs rather than straight lines describe these distances. The angular size of an arc is measured by the angle formed by the radial lines that connect the center of the sphere to each of the endpoints of the arc. Thus, the angular size of the arc **AB** in the figure is the angle **AOB** and is labeled **θ.** Similarly, the angular distance of the star at point **C** from the equator along the star's meridian is the angle **BOC** and is labeled **φ** in the figure. Thus, any location on the sphere can be specified uniquely by using these two coordinates, **θ** and **φ.**

A familiar example of a spherical coordinate system is that of latitude and longitude on the surface of the Earth.

1. Launch *Starry Night Enthusiast*™ and select **Favourites > Observing Projects > Coordinate Systems > Latitude and Longitude** from the menu.

Starry Night Enthusiast™ shows a view of Earth as seen from space. The viewing location is 12,000 kilometers above the surface of the Earth along a line connecting the center of the Earth and the center of the Sun. The image of the Earth is overlaid with a grid representing the terrestrial spherical coordinate system of **latitude** and **longitude.** The equator of the coordinate system is the red line running horizontally across the center of the Earth. The reference plane of the equator is chosen to be perpendicular to the rotation axis of the Earth. In this way, the north and south poles of the coordinate system, indicated by the short blue and yellow lines extending from the top and bottom of the Earth's surface in the view, coincide with this rotation axis.

Latitude is the angular distance north or south of the equator. The view shows several parallels of latitude, so-called because they are parallel to the equator. Notice that they are labeled in units of degrees, positive to the north of the equator and negative to the south of the equator.

The vertical red line across the surface of the Earth is the reference meridian for longitude. This reference meridian, also called the **prime meridian,** was chosen earlier in history to pass through the site of the Old Royal Observatory at Greenwich, England. Longitude is measured east or west of this prime meridian. The two coordinates of latitude and longitude uniquely specify any location on the surface of the Earth.

Question 1. Are the meridians of longitude great circles? Why or why not?

Question 2. Are the parallels of latitude great circles? Why or why not?

Question 3. Which of the two coordinate angles depicted in Figure 1, (θ or φ), corresponds to
a) latitude, and b) longitude?

2. Position the mouse cursor over the image of the Earth and use the **Location Scroller** to rotate the Earth in the view in order to look nearly straight down onto the north pole (indicated by the blue pole stick). Note that the meridians of longitude converge at the pole. Now use the **Location Scroller** to rotate the Earth so that the view is nearly straight down onto the south pole (yellow pole stick). Notice again that the meridians of longitude all intersect at the pole.

You may have noticed that *Starry Night Enthusiast*™ has labeled the meridians of longitude in units of time (hours), rather than units of angle (degrees). The reason for this is that the Earth rotates once on its axis in 24 hours with reference to the direction to the Sun. Consequently, any location on the Earth moves 360° around a circle oriented parallel to the equator in 1 day of 24 hours. Thus, longitude can also be expressed in units of time, 1 hour of longitude being equal to the angle turned through by the Earth, in that time; that is, 360 ÷ 24 degrees.

3. Select **File > Revert** from the menu. Notice that from this perspective the prime meridian lines up with the poles.

4. Change the **Time Flow Rate** to **3000×.**

As time passes, note that the meridians of longitude shift along with the Earth as the Earth rotates toward the east. In other words, the coordinate system is fixed on the Earth.

Question 4. How, if at all, does the passage of time affect the latitude and longitude of a location upon the surface of the Earth?

5. Select **File > Revert.**

6. Change the **Time Flow Rate** to **2 hours** and **Step time forward.** Notice that with each step forward in time, the meridians of longitude shift along with the Earth toward the east so that a different meridian lines up with the poles as seen from this vantage point in space.

7. Continue to **Step time forward** until the prime meridian once again lines up with the poles.

Question 5. How many hours elapse until the prime meridian is once again lined up with the poles in this view?

Question 6. What is the angular distance in degrees between the meridians of longitude shown in the view?

Question 7. What is the approximate longitude of the location marked **Messina** in southern Africa, expressed in degrees?

B. Horizontal Coordinates

The previous section introduced a spherical coordinate system centered on the Earth. This section of the project will explore a spherical coordinate system for the celestial sphere from the perspective of an observer on the surface of the Earth.

The most natural reference plane for this observer is the plane that is tangent to the point of the observer's location on the surface. This plane intersects the celestial sphere in a great circle at the **horizon** and is the equator of the horizontal coordinate system. The sky occupies the hemisphere above the horizon. The point directly overhead, 90° from the horizon, is the center of this hemisphere. It is one of the poles of the coordinate system and is known as the **zenith.** The center of the hemisphere beneath the horizon, antipodal to the zenith and hidden by the Earth is the **nadir.**

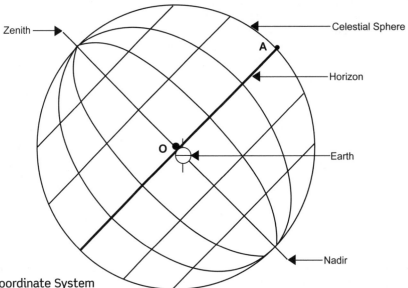

Figure 2. Horizontal Coordinate System

Figure 2 shows the horizontal coordinate grid for an observer at a mid-northern latitude on the Earth at the point marked **O**, as seen from a position outside the celestial sphere. The point marked **A** indicates the north point on the observer's horizon, the chosen reference point on the equator of the horizontal coordinate system.

In the horizontal coordinate system, the position of an object's meridian along the horizon measured eastward from the north point is its **azimuth.** The great circle that passes through the zenith and the north point of the horizon is called the **local** or **celestial meridian.**

> The meridian is illustrated in Figure 2-21 of Freedman, Geller, and Kaufmann, *Universe*, 9th Ed.

The position of an object above (or below) the horizon along its meridian is its **altitude,** also measured in degrees. Altitude is positive for objects above the horizon and negative for objects below the horizon. With these two coordinates, azimuth and altitude, you can specify any location on the celestial sphere.

You can use *Starry Night Enthusiast*™ to view the sky from the center of this coordinate system and use the coordinates of azimuth and altitude to specify the positions of several objects.

8. Click the **Home** button.

Starry Night Enthusiast™ should show a view of the horizon and sky looking south from your home location at the current time and date. Compass points are indicated along the horizon. In addition, the **Gaze** panel on the right of the toolbar displays the horizontal coordinates of the center point in the view.

9. Click on the **Gaze** buttons in the toolbar (**N, E, S, W,** and **Z**) to look around the view and use the **Gaze** display panel to help you to answer the following questions.

Question 8. What is the azimuth of each of the compass points a) north, b) east, c) south, and d) west?

Question 9. What is the altitude of the zenith?

Question 10. What is the azimuth of the zenith? [*Hint:* Does the north pole of Earth have a specific longitude?]

10. Click **Home** and then select **View > Local Meridian**. *Starry Night Enthusiast*™ displays the local meridian for your current viewing location. Notice that it is marked in degrees.

11. Select **View > Zenith/Nadir** from the menu.

12. Use the **Gaze** buttons to look around the view.

13. Open the **Options** pane and remove the horizon from the view by clicking the checkbox to the left of the label **Local Horizon** under the **Local View** layer.

14. Use the **Gaze** buttons to look around the view.

15. Press the **N** button to change the gaze to due north and then use the **Hand Tool** to drag the view upward on the screen so that your gaze is directed straight down toward the nadir.

> **Question 11.** In addition to the zenith and the north point on the compass, what other specific points are included on the great circle described by the local meridian?

> **Question 12.** What is the altitude of the nadir?

16. Open the **Options** pane and double-click on the **Local Horizon** label to open the **Local Horizon Options** dialog box. Click the **Select Horizon Panorama...** button. In the dialog box that pops up, click on the **Flat Grass** option and then click the **Select** button. Finally, click the **OK** button in the **Local Horizon Options** dialog box.

17. In the **Options** pane, click the checkbox next to the **Local Horizon** label to turn the display of the horizon back on.

18. Before closing the **Options** pane, click the checkbox next to the **Local Grid** label in the **Guides** layer.

19. Use the **Gaze** buttons to look around the view.

Starry Night Enthusiast™ displays a grid of horizontal coordinates. Parallel to the horizon line are circles of altitude. Perpendicular to the horizon are meridians of azimuth. Note that the meridians of azimuth converge as you look toward the zenith.

> **Question 13.** Are the circles of altitude great circles? Why or why not?

> **Question 14.** Are the meridians of azimuth great circles? Why or why not?

> **Question 15.** Which of the two coordinate angles depicted in Figure 1 (θ and ϕ) corresponds to a) altitude, and b) azimuth?

20. Select **Favourites > Observing Projects > Coordinate Systems > Altitude-Azimuth** from the menu.

Starry Night Enthusiast™ shows a view of the horizon and sky looking south from Munich, Germany, at noon on January 22, 2006. The horizontal coordinate system is displayed and the Sun and Venus are labeled.

> **Question 16.** Using the local grid as a guide, what is your estimate of the altitudes and azimuths of a) the Sun, and b) Venus from this location at this time and date?

21. In the **File** menu (the *Starry Night Enthusiast*™ menu on a Macintosh), open the **Preferences** dialog and select **Cursor Tracking (HUD)** from the drop box. In the **Show** list, select **Altitude, Azimuth,** and **Name.**

22. Position the **Hand Tool** over the Sun and use the **HUD** to find its precise altitude and azimuth.

23. Position the **Hand Tool** over Venus and note its exact altitude and azimuth from the display in the **HUD.**

Question 17. As seen from Munich, Germany, at noon on January 22, 2006, what are the precise altitudes and azimuths of a) the Sun, and b) Venus?

A disadvantage of the horizontal coordinate system is that the coordinates of an object vary with the observer's location on the surface of the Earth because the celestial sphere is centered on the observer.

24. Select **Options > Viewing Location…** from the menu and select **Algiers, Algeria,** from the list of locations. Click the **Set Location** and then press the **spacebar** to move to the new location.

25. Set the **Gaze** to the south, and use the **HUD** to determine the horizontal coordinates of Venus and the Sun as seen from Algiers at noon on January 22, 2006.

Question 18. As seen from Algiers at noon on January 22, 2006, what are the horizontal coordinates, azimuth and altitude of a) the Sun, and b) Venus?

The coordinates of an object in the horizontal coordinate system vary with the observer's location. Horizontal coordinates of objects in the sky also vary with time as the observer's location is carried eastward by the Earth's rotation over time. This rotation of the Earth shifts the orientation of the reference plane in space. Figure 3 shows how the coordinate system for the observer at the location marked **O** in Figure 2 appears 12 hours later when the Earth's rotation has carried the location through 180°.

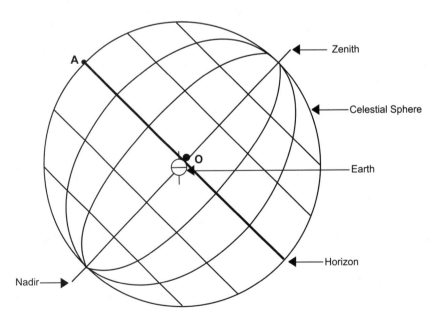

Figure 3. Horizontal Coordinate System 12 Hours Later

26. Set the **Time** in the toolbar to **2:00:00 PM.** Use the **HUD** to determine the horizontal coordinates of the Sun and of Venus.

> **Question 19.** As seen from Algiers at 2:00:00 PM on January 22, 2006, what are the horizontal coordinates of a) the Sun, and b) Venus?

C. Equatorial Coordinates

The horizontal coordinate system, discussed above, places the observer's location on the surface of the Earth at the center of the celestial sphere and defines a reference plane that is tangent to the Earth's surface at this location. Thus, the horizontal coordinates of objects in the sky are specific to the observer's location and local time and are inconvenient when attempting to communicate an object's position to someone at another location.

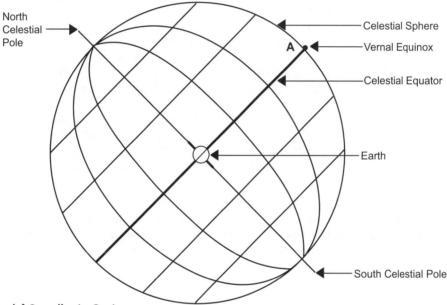

Figure 4. Equatorial Coordinate System

It is desirable to define a coordinate system in which the coordinates of objects in the sky do not vary because of changes in the observer's location or by the passage of time. In practice, this means defining a reference frame that is fixed with respect to the celestial sphere rather than one which is associated with a specific observer's position. Figure 4, above, depicts the **equatorial coordinate system,** as seen from beyond the celestial sphere. In this coordinate system, the center of the celestial sphere coincides with the center of the Earth. The reference plane is an extension of the plane of the Earth's equator, hence the name for this coordinate system. This reference plane intersects the celestial sphere in a great circle called the **celestial equator.** The line perpendicular to this plane and passing through the center of the Earth thus coincides with the rotation axis of the Earth and points to the **north** and **south celestial poles** of the coordinate system.

In the equatorial coordinate system, the perpendicular angular distance of an object from the celestial equator is called **declination.** Declination is similar to altitude in the horizontal system and is measured in degrees and is positive north of the celestial equator and negative south of the celestial equator.

The coordinate denoting position around the celestial equator is called **right ascension** and is usually measured in units of time. The reason for this is that the vernal equinox moves through 360° in 24 sidereal hours. The reference point from which an object's right ascension is measured is a specific point known as the **vernal equinox** (marked **A** in Figure 4). The vernal equinox is the point in the sky where the Sun crosses the celestial equator as it moves northward in the sky. Thus, the Sun is at this position on the first day of spring in the northern hemisphere of the Earth every year. Right ascension increases in an eastward direction from the vernal equinox.

Question 20. Which of the two coordinate angles depicted in Figure 1 (θ or φ) corresponds to a) declination, and b) right ascension?

27. Select **Favourites > Observing Projects > Coordinate Systems > Coordinates From North Pole.**

Starry Night Enthusiast™ shows a view of the sky from the north pole of Earth. The two coordinate systems are overlaid on the view, with the horizontal coordinate system for this location shown in grey and the equatorial coordinate system in red. The location of the north pole provides a special perspective from which to compare the two coordinate systems. Note that the horizon and the celestial equator are coincident from this observing location.

Recall that in the horizontal coordinate system, the celestial sphere is centered on the observer's location on the surface of the Earth. The center of the celestial sphere in the equatorial coordinate system is at the center of the Earth. At the north pole, the plane of the horizon (the reference plane in the horizontal coordinate system) is parallel to the plane of the Earth's equator (the reference plane in the equatorial coordinate system) and the zenith coincides with the north celestial pole. (In fact, the center of the celestial sphere shifts by a distance equal to the radius of the Earth between the two coordinate systems, but this distance is negligible compared to the enormous distance to the surface of the celestial sphere. Thus, the coordinate systems are closely equivalent and parallels of altitude and parallels of declination are coincident in the view from this location.)

This view shows the position in the sky of the vernal equinox, the reference point for the equatorial coordinate system. The great circle in this coordinate system that passes through the vernal equinox is the zero point of right ascension. The grey labels along the celestial equator (which is equivalent to the horizon at this location) indicate the right ascension of other great circles in the equatorial coordinate grid.

28. Use the **Gaze** buttons to look around the view. When you gaze at the zenith, open the **Options** pane and turn off the **Celestial Poles** option and turn on the **Zenith/Nadir** option. Note that from the north pole of Earth, the zenith and the north celestial pole are coincident at this location on Earth.

Question 21. a) What is the altitude of the north celestial pole for an observer at the north pole of the Earth? b) What is the declination of the north celestial pole? c) Does your answer to part b depend on your observing location on the Earth? Why or why not?

29. Close the **Options** pane. Use the **Hand Tool** to drag the view slightly in any direction and then select **File > Revert.**
30. **Run time forward.** Observe the relative motion of the two coordinate grids.
31. Select **Edit > Undo Time Flow** to return to the initial view.
32. Change the **Gaze** to the **Zenith.**
33. **Run time forward.** Observe the relative motion of the two coordinate grids.

Question 22. How do you explain the relative motion between the equatorial coordinate system and the horizontal coordinate system as time passes for an observer on the Earth? (*Hint:* Consider what effect the rotation of the Earth will have on each of these coordinate systems.)

Question 23. How do the stars and other objects in the sky move relative to the two coordinate system grids as time progresses?

34. Select **Favourites > Observing Projects > Coordinate Systems > Coordinates From Mid-Northern Latitude.**

At the left of the view, you will notice a label indicating the star **Hamal.** In the next sequence of steps, you will use *Starry Night Enthusiast*™ to observe and compare the altitude and azimuth coordinates of this star with its right ascension and declination as seen from a mid-northern latitude of Earth.

35. Open the **Info** pane. Under the layer headed **Position in Sky,** the horizontal coordinates of altitude and azimuth, as well as the equatorial coordinates of right ascension and declination, for the star **Hamal** are shown.

36. **Run time forward.** As time progresses, observe how the passage of time affects the coordinates of this star.

37. Select **Edit > Undo Time Flow** and note the local and equatorial coordinates of the star **Hamal** from the **Info** pane.

38. Select **Favourites > Observing Projects > Coordinate Systems > Coordinates from North Pole** again and note the local and equatorial coordinates of the star **Hamal** from this location. Select **Edit > Undo "Favourite" Item** and **Edit > Redo "Favourite" Item** to see the coordinates of this star at each location again.

Question 24. Is either coordinate system independent of the observer's position; in other words, does either coordinate system designate coordinates that remain constant from location to location?

Question 25. Is either coordinate system independent of the passage of time; in other words, does either coordinate system designate coordinates that remain consistent over time?

Question 26. If you wanted to produce an enduring map of the sky with coordinates for the objects that appear on the celestial sphere, which coordinate system would you choose?

Question 27. If you and a friend were observing the sky and you wanted to indicate the location of a particular star to your friend, which coordinate system would be most useful?

Question 28. If you wanted a fellow astronomer in another city to observe a specific object in the sky, what coordinates would you provide to your colleague?

D. Other Coordinate Systems

Several other astronomical coordinate systems have been designed for specific purposes. The first of these, the **ecliptic coordinate system,** is used mainly for observations of objects in the solar system. In this system, the celestial sphere is centered on the center of the Earth and the plane of the Earth's orbit around the Sun, the ecliptic plane, is used as the reference plane. This plane intersects the celestial sphere at the ecliptic equator and the reference point on the ecliptic equator is the vernal equinox. The poles of the ecliptic coordinate system are the north and south ecliptic poles. The coordinates of the ecliptic reference frame are ecliptic latitude and ecliptic longitude.

39. Select **Favourites > Observing Projects > Coordinate Systems > Ecliptic Coordinates** from the menu and use the **Gaze** buttons and **Time Flow** controls to see how the sky moves against this coordinate system.

Question 29. Do the positions of the stars move relative to the ecliptic coordinate system as time progresses?

Another astronomical coordinate system is the **galactic coordinate system,** which is used in studies of our own Galaxy. Again, the celestial sphere is centered on the center of the Earth but now the reference plane is the mean plane of the Milky Way, which intersects the celestial sphere at the galactic equator. Galactic latitude is measured north and south of the galactic equator. The poles of the system are the north and south galactic poles. The reference point on the galactic equator is the direction to the galactic center and galactic longitude is measured eastward from this point in the sky along the galactic equator.

40. Select **Favourites > Observing Projects > Coordinate Systems > Galactic Coordinates** from the menu and use the **Gaze** buttons and **Time Flow** controls to see how the sky moves against this coordinate system.
41. You can show the relationship of this coordinate system to the Milky Way Galaxy by displaying it in the view by clicking on **View > Stars > Milky Way.** You can brighten this galaxy by using the slidebar in **Options > Stars > Milky Way.**

Question 30. Do the positions of the stars move relative to the galactic coordinate system as time progresses?

42. Do not save any of the changes you have might have made to the views when you exit *Starry Night Enthusiast*™.

E. Conclusion

In this project, you have learned about spherical geometry and how we apply this geometry to various coordinate systems that are used by astronomers to map the sky. In practice, the celestial coordinate system universally used by astronomers is not fixed with respect to the background sky over long time scales because of the precession of the spinning Earth and the slow drift of the equator and poles that are directly related this motion. Thus, positions on this coordinate system are assigned a specific date and astronomers must adjust coordinates for this precessional motion.

Changing Latitude 4

The ancient Greeks knew that they would see stars above the southern horizon that could not be seen from Greece if they traveled south. Conversely, when they traveled north, stars that they could see in Greece no longer rose above the southern horizon.

The Greeks realized that this is exactly what would happen if the Earth were round. On a flat Earth, the same stars should be above the horizon for all observers, regardless of where they were located. On a round Earth, on the other hand, our view of the sky changes when we move to different latitudes.

The effect of changing latitude on our view of the sky is discussed in Section 2–4 and illustrated in Figure 2–11 of Freedman, Geller, and Kaufmann, *Universe*, 9th Ed.

In this observing project, you will learn how our view of the stars and constellations changes as we move to different latitudes in the northern hemisphere.

A. The View to the South

1. Launch *Starry Night Enthusiast*™ and select **Favourites > Observing Projects > Changing Latitude > Mexico City** from the menu.

2. Select **Preferences** from the **File** menu (Windows) or **Starry Night Enthusiast** menu (Macintosh). Choose **Cursor Tracking (HUD)** in the upper left drop box and select **Altitude, Name,** and **Object Type** from the **Show** list. Check the option to display **HUD** data in the upper left corner of the screen in the **Preferences** dialog if you prefer. These chosen parameters will now appear when the cursor is placed over any object.

3. Close the **Preferences** dialog window.

This view shows the sky looking south from Mexico City, Mexico, (latitude 19° N) at 1:29:22 AM Standard Time on April 24, 2007.

The constellation Scorpius appears just above and to the left of center in the view. The top end of Scorpius in this view is shaped like a letter T tilted over on its right side, and the bottom end is a tail shaped like a fishhook. The bright red star is Antares, a cool but very bright supergiant star that can be used as an aid to the identification of Scorpius in the real sky. The other labeled star at the center of the T is Dschubba, which we shall use as a reference star in the next series of observations.

4. Check your identification of Scorpius by opening the **View > Constellations** menu and selecting both the **Labels** and **Astronomical** options.

5. Select **View > Local Meridian** from the menu.

Note that the reference star, Dschubba, is due south in this view, transiting the meridian at this precise time. Therefore, you are seeing it at its highest position in your sky. This occurs because the stars and constellations rise in the eastern part of the sky and set in the western part, just as the Sun does. They are highest in the sky when they are half way between rising and setting; i.e., when they are due south as seen from the northern hemisphere. You can see that the tail of the scorpion extends across the Milky Way toward the horizon.

6. Position the **Hand Tool** over Dschubba and note its **Altitude** in the **HUD**. Record this value in Data Table 1.

7. Open the **Status** pane and expand the **Location** layer. Note the **Latitude** of Mexico City in Data Table 1. Then close the **Status** pane.

In order to see how our view of the sky changes with latitude, we will move progressively northward and note the altitude of Dschubba when it transits the meridian at each new location.

8. Select **Favourites > Observing Projects > Changing Latitude > Chicago** from the menu.

9. Position the **Hand Tool** over Dschubba and note its **Altitude** as seen from Chicago. Record this value in Data Table 1. Obtain the **Latitude** of Chicago from the **Location** layer in the **Status** pane, as before, and record this value in Data Table 1.

10. To answer the following questions, you may want to switch back and forth between the views from Mexico City and Chicago. To do so, alternately select **Edit > Undo "Favourite" Item** and **Edit > Redo "Favourite" Item** from the menu.

Question 1. In which direction does the constellation Scorpius move with respect to the south horizon as the viewing location moves north from Mexico City to Chicago?

Question 2. Why is the time at which Dschubba transits the meridian different for the two locations?

11. Select **Favourites > Observing Projects > Changing Latitude > Regina** from the menu. You can see that the **Date** and **Time** have changed in order to place Dschubba on the meridian at this new location.

12. Position the **Hand Tool** over Dschubba and note its **Altitude** at the time it transits the meridian at Regina, Canada. Record this value and the **Latitude** of Regina from the **Location** layer of the **Status** pane in Data Table 1.

Question 3. Where is Scorpius, with respect to the horizon on the screen, as seen from Regina? Is any part of Scorpius missing? (You can display the shape of this constellation as a guide by checking **View > Constellations > Astronomical**).

13. Select **Favourites > Observing Projects > Changing Latitude > Fairbanks** from the menu and note the **Altitude** of Dschubba and the **Latitude** of Fairbanks in Data Table 1.

Data Table 1. Altitude of Dschubba Transit from Cities at Various Northern Latitudes

Location	Latitude	Altitude of Dschubba
Mexico City, Mexico		
Chicago, USA		
Regina, Canada		
Fairbanks, USA		

14. To help answer the following questions, open the **Favourites** pane and select the files **Mexico, Chicago, Regina,** and **Fairbanks** in turn under the **Observing Projects > Changing Latitude** folder. You may also choose to select **View > Constellations > Astronomical** in each of the views.

Question 4. Where is Scorpius on the view as seen from Fairbanks? What part of Scorpius is visible? Where is the rest of Scorpius, relative to the horizon?

Question 5. For which of the cities in Data Table 1 is Scorpius completely above the horizon? (We are referring to the "constellation" here in the historical sense, as the pattern of bright stars shaped like a T and fishhook, rather than as the entire astronomical area of sky.)

Question 6. At about what latitude does the tail of Scorpius begin to disappear from view (i.e., NEVER rises above the horizon) for an observer traveling north?

Question 7. Use the observed positions of Scorpius on the screen to estimate the approximate latitude above which NO part of Scorpius is ever visible.

Question 8. Which of the following describes the relationship between latitude and the elevation angle between the reference star and the horizon?

a) As latitude increases, the reference star's elevation angle increases.

b) There is no relationship between the reference star's elevation angle and latitude.

c) As latitude increases, the reference star's elevation angle decreases.

Question 9. Suppose that a scientist in Fairbanks has applied for a research grant to study a globular cluster of stars near the middle of Scorpius, using an observatory just outside Fairbanks. If you were on the granting agency, what would be your response to this proposal? Would your response change if the research proposal included a request for funds to travel to an observatory in Texas?

Question 10. In the data recorded in Data Table 1 above, of altitude A of the star Dschubba when measured at sites with latitude λ, which of the following mathematical relationship describes these data?

a) $\lambda - A = $ Constant

b) $A = 90° - \lambda$

c) $\lambda + A = $ Constant

d) $A = $ Constant $\times \lambda$

The Appendix at the end of this project shows the derivation of the relationship between altitude of a star and its declination and the latitude of the observing site. Declination, one of the equatorial coordinates of a star, is defined in the Coordinate Systems chapter earlier in this book.

B. Pole Star Altitude and Latitude on the Earth

> 15. Select **Favourites > Observing Projects > Changing Latitude > Minneapolis** from the menu.

The view is now northward from Minneapolis, Minnesota, at 7:30 PM standard time on October 1, 2005. Polaris, the North Star, is labeled. **Time Flow** is frozen and the **Time Step** is set at 3000 times the rate of real time. Identify the Big Dipper in the constellation Ursa Major on the left of the view and Cassiopeia on the upper right.

> 16. **Run time forward** and watch what happens in the sky.

> **Question 11.** What happens to stars on the left side of the view, including the Big Dipper, as time progresses? E.g., do they move upward or downward?

> **Question 12.** What happens to stars on the right side of the view, including Cassiopeia, as time progresses? E.g., do they move upward or downward?

> **Question 13.** Is there one star that appears to remain at rest while all other stars move in circles around it? Which star is this?

As the sky rotates around the pole, watch what happens to the Big Dipper, especially between about 10 PM and 1 AM. In the lower left side of the view, stars are setting below the horizon, and on the lower right side of the view, stars are rising above the horizon. From Minneapolis, however, the Big Dipper never sets. It approaches the northern horizon, but passes above this horizon without setting, and then gets higher in the sky again.

At about 1:30 AM, as the Big Dipper is starting to rise again, you should see a very bright star move into the view on the left. This is the second brightest star in the northern hemisphere, **Vega**, a bright, blue main-sequence star in the constellation **Lyra**, the Lyre.

> **Question 14.** What happens to Vega at about 3:45 AM?

Circumpolar stars are described in Section 2–4 of Freedman, Geller, and Kaufmann, *Universe*, 9th Ed.

Stars or constellations that never set as they move in circles around the pole are called **circumpolar**. The Big Dipper is circumpolar as seen from Minneapolis, while Vega is not.

> 17. Select **File > Revert** from the menu to return to the initial view.
> 18. Position the cursor over Polaris and, from the **HUD**, note its **Altitude** above the northern horizon. Record this value in Data Table 2.
> 19. From the **Status** pane, note the **Latitude** of Minneapolis in Data Table 2.

Question 15. How does the elevation angle of the Pole Star, Polaris, compare to the latitude of Minneapolis? (These numbers should be close but will not be precisely equal because the Pole Star is actually about one degree away from the position of the north celestial pole in the sky.)

20. Select **Favourites > Observing Projects > Changing Latitude > Houston** from the menu.

The view is to the north from Houston, Texas, at the same time and date as the previous view from Minneapolis.

21. Note the **Altitude** of Polaris above the northern horizon from Houston and record this value in Data Table 2. From the **Status** pane, find the **Latitude** of Houston and enter this value in Data Table 2 as well.

Data Table 2. Latitude and Pole Star Altitude from Minneapolis and Houston

Location	Latitude	Pole Star Altitude
Minneapolis		
Houston		

Question 16. How does the elevation angle of the Pole Star from this location compare with the latitude of Houston?

22. **Run time forward** and watch what happens in the sky.

Notice that all of the stars of the Big Dipper except the one at the top right corner of the bowl (the one closest to the Pole Star) set below the horizon, then rise again later.

Question 17. Out of all the stars in the sky, how does the number that is circumpolar as seen from Houston compare with the number as seen from Minneapolis (e.g., are more stars circumpolar as seen from Houston, less, etc.)?

23. Select **Favourites > Observing Projects > Changing Latitude > Equator** from the menu.
24. **Run time forward.**

Question 18. Looking north from the equator, are any stars circumpolar?

Question 19. Is Polaris circumpolar as seen from the equator? Why or why not?

25. Select **Favourites > Observing Projects > Changing Latitude > North Pole** from the menu.

26. **Run time forward**. As time flows, use the **Hand Tool** to scroll the view eastward through a full 360°.

27. Click the **Z** button in the **Gaze** section of the toolbar to see Polaris near the zenith.

Question 20. From the north pole, are there any stars that are *not* circumpolar?

Question 21. Based on your observations, which one of the following statements do you think is correct?

    a) The angle of the Pole Star above the horizon equals your latitude.

    b) The angle of the Pole Star above the horizon equals 90° minus your latitude.

    c) The angle of the Pole Star above the horizon does not depend on your latitude.

Question 22. If you were south of the equator, say in Australia, would you expect to see circumpolar constellations anywhere in the sky? If so, in what part of the sky would they be?

C. Simple Celestial Navigation

Imagine that you have been selected by the National Geographic Society to lead an expedition recreating Columbus's historic 1492 voyage to the New World. Your only navigational aids are a compass and a quadrant, a device used to measure the elevation angle (altitude) of stars and planets.

Like Columbus, you leave Palos de la Frontera, Spain, on August 3 and use dead reckoning and coastal navigation to reach La Gomera, Spain, in the Canary Islands. There you take on supplies and effect any necessary repairs to your ships.

On September 6, you set sail from La Gomera (latitude 28° 6' N) to cross the Atlantic Ocean, intending to meet the sponsors of the expedition in Nassau, Bahamas, (latitude 25° 3' N) some time in October.

28. Close the **Status** pane if it is open.

29. Select **Favourites > Observing Projects > Changing Latitude > La Gomera** from the menu.

30. **Run time forward** and observe the night sky seeking stars and constellations that will aid you in your navigation. Locate the Pole Star and note its altitude.

You wish to maintain a heading slightly south of due west without straying too far north or south. Unfortunately, four days into the voyage, a five-day storm blows you off course, leaving you lost. Nine days after leaving La Gomera, the night sky is finally clear. You need to determine whether to correct your subsequent heading toward the north or south in order to reach Nassau.

31. Select **Favourites > Observing Projects > Changing Latitude > Lost** from the menu.

32. **Run time forward** and observe the sky through the night.

Question 23. To get back on course, do you need to adjust your heading to the south or north of west after surviving the storm?

D. Eastern Rising of the Sun in the Sky

It is interesting to explore the path of the Sun as it rises above the eastern horizon and to measure the angle that this path makes with the horizon from various latitudes. There is a great difference between sunrises (and in an equivalent manner, sunsets) when they are observed from different latitudes. For example, at the equator, the Sun appears rather suddenly and rises rapidly into the sky after only a short period of twilight. In contrast, at high latitudes, the Sun moves more slowly across the horizon after a prolonged period of twilight. At very high latitudes during the summer, the Sun is always above the horizon.

In this section of the project, you will measure the angle between the Sun's track and the horizon, which we can call the "rising angle," and investigate the dependence of this angle on the latitude of the observing site.

33. Select **Favourites > Observing Projects > Changing Latitude > Seattle Sunrise.**

The view shows the upper limb of the Sun just rising in the East as seen from Seattle on March 19, 2006. This date, the first day of spring in the northern hemisphere (and the first day of fall or autumn in the southern hemisphere!), has been selected to place the Sun very close to due east at sunrise as a convenient reference point. The **Time Step** is set to one hour.

34. **Step time forward** one hour and note the direction in which the Sun has moved.

The Sun will have moved upward and to the side during this first hour after sunrise.

Question 24. In which direction has the Sun moved?

The geometry of the Sun's position one hour after sunrise is shown in Figure 1. In that hour, the Sun has moved some distance above the horizon and some distance away from the east point on the horizon. The distance that the Sun has moved above the horizon in the interval of one hour is the shortest distance between the Sun and the horizon. This distance will be on the line from the Sun that intersects the horizon at a right angle. We will call the point of this intersection H so that the distance of the Sun above the horizon is SH. As you can see from the figure, the distance from the Sun to the east point of the horizon, which we will call SE, is the hypotenuse of a right triangle. (In practice, the track of the Sun across the sky will be slightly curved and the use of its position one hour after sunrise will underestimate the true rising angle, but this method is sufficiently accurate for our purposes.)

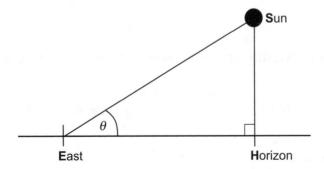

Figure 1. Rising Angle, q, of the Sun

You can measure two angular separations, SH and SE, on the sky and these can be assumed for this purpose to be distances. The ratio of these two "distances," SH/SE, is known as the sine trigonometric function of the rising angle, θ, and is written as Sin θ. The inverse sine of this value gives the rising angle, θ.

Sequence 1

35. Open the **Status** pane and expand the **Location** layer. Note the **Latitude** of the current viewing location and record it in the appropriate column of Data Table 2.

36. Use the **Angular Measurement Tool** to measure the angular distance from the Sun to the point on the horizon directly below the Sun. Note the measurement in degrees and arcminutes, to the nearest arcminute, in Data Table 2 under the heading Measured SH. Convert this measurement to decimal degrees by dividing the number of arcminutes by 60 and adding this quotient to the number of degrees in the measurement. Write this value in the column labeled SH in Data Table 2.

37. Next, measure the angular distance from the Sun to the East point of the horizon. Note this measurement in degrees and minutes of arc to the nearest arcminute in Data Table 2 under the column labeled Measured SE. Then convert this measurement to decimal degrees and write the result in the column labeled SE in Data Table 2.

38. Calculate the ratio SH/SE and enter the value in Data Table 2 under the column labeled Sin θ.

39. Use the inverse sine function on a calculator or spreadsheet to determine the angle, θ, in degrees that corresponds to the calculated value of Sin θ.

Data Table 3. Rising Angle of the Sun from Locations at Different Latitudes

LOCATION	Latitude °	Measured SH °		'	SH °	Measured SE °		'	SE °	Sin θ = SH/SE	θ °
Seattle											
Honolulu											
Quito											
Dunedin											
Murmansk											
North Pole											

E. Dependence of the Sun's Rising Angle on the Observer's Latitude

It is interesting to explore the dependence of this "rising angle" of the Sun on the latitude of the viewing location by repeating the above procedure for the other locations on the Earth that are listed in Data Table 2.

40. For each location listed in Data Table 2, select the appropriately named view from **Favourites > Observing Projects > Changing Latitude >** *location* **Sunrise** and repeat the procedure outlined in Sequence 1 above.

The position and time of sunrise will vary somewhat between sites because of time zone differences. The situation at the north pole is interesting. You will have to consider what is the "rising angle" if the Sun just skims across the horizon!

When you have entered data for each location in Data Table 2, try answering the following questions.

Question 25. At what angle from the horizon does the Sun rise for someone at 0° latitude?

Question 26. At what angle from the horizon does the Sun rise for someone at 90° latitude?

Question 27. How does the rising angle of the Sun change as you move from Quito, Ecuador (0° latitude) to Honolulu, then to Seattle, Murmansk, and finally to the north pole (90° latitude)?

Question 28. Based on your answers to the questions above, which ONE of the following statements do you think is correct?

a) The rising angle of the Sun at your location is equal to your latitude.

b) The rising angle of the Sun at your location is equal to 90°minus your latitude.

c) The rising angle of the Sun does not depend on your latitude.

Question 29. What is different about the sunrise in Dunedin, New Zealand, compared with that in Seattle? [Hint: Think about the direction in which the Sun rises from the horizon.]

Question 30. Suppose that you are in Dunedin, New Zealand, the time is noon, and you are facing east, toward the point on the horizon where the Sun rises. Based on the direction toward which the Sun rises in Dunedin, in which direction (left or right) would you need to turn to face the Sun at noon in Dunedin? How does this compare to Seattle?

Question 31. Based on how the Sun rises in Dunedin, in which compass direction (north or south) would you need to face to see the Sun at noon? How does this compare to Seattle?

Question 32. Based on how the Sun rises in Dunedin, in which direction would the Sun be moving if you were faced toward it at noon (from left to right or from right to left)? How does this compare to what you saw in Seattle at noon?

41. Check your answer to the previous question directly by selecting **Favourites > Observing Projects > Changing Latitude > Dunedin Sunrise** and adjusting the time in the toolbar to **12:00:00** PM (noon).

42. Change the **Gaze** to the North.

43. Set the **Time Flow Rate** to **1 minute**, and **Run time forward**.

Question 33. Toward which compass direction would the Sun be moving at noon in Dunedin (from east to west or from west to east)? Check the compass directions near the bottom of the view to find if you are right. How does this compare to what you would see in Seattle?

F. The Rising of Objects at Night

We have observed the rising of the Sun in both hemispheres. It is interesting to verify that other objects, when observed at night, also show the same motion and geometry as they pass over the horizon. Set up the following initial conditions and observe the eastern horizon as time advances.

44. Select **Favourites > Observing Projects > Changing Latitude > Seattle Sunrise** from the menu.

45. Select **View > Hide Daylight**.

46. Set the **Time Flow Rate** to **300×**. Allow time to advance for a few hours and compare the rising angle of the stars with that of the Sun.

47. Now set the **Gaze** to due **West**. Allow time to continue to flow and use the **Gaze** controls in the toolbar to look alternately **East** and **West** and compare the angle that the stars make with the eastern horizon as they rise with the angle they make with the western horizon as they set, their setting angles.

Question 34. How does the rising angle of the stars compare with that of the Sun?

Question 35. How do the rising and setting angles of the stars compare?

48. Close *Starry Night Enthusiast™* without saving changes to the file.

G. Conclusions

In this project, you have seen that the view of the southern sky becomes more and more restricted as you move northward in the northern hemisphere until, near to the poles, only about one-half of the celestial sphere is visible. Related to this, you have seen that more and more of the stars on this celestial sphere are circumpolar, that is, they can be seen throughout the night from any observing site.

Of course, the same would be true in the southern hemisphere as one moved further south toward the southern polar regions until, from the south pole, only one-half of all stars would be visible. However, from this site, these would be in the opposite hemisphere from the stars visible from the north pole.

The other major fact that you have verified is that the angle between the horizon and the north celestial pole, approximated at this time in history by the Pole Star, is equal to the latitude of the observing site. You had the opportunity to discover how early navigators made use of this fact.

You evaluated the relationship between the observer's latitude and the angle at which objects rise and set relative to the horizon and observed the difference in sunrises between the two hemispheres and unusual sunrises at the north pole.

Appendix: The Geometry of the Relation Between Altitude and Latitude

In this project, you have demonstrated two relationships between the latitude λ of the observing site and the altitude of a star. The first of these examined the variation of the angle above an observer's horizon of a fixed star as latitude changed. The second looked at the same variation of angle above the horizon of a specific, and easily recognizable, position in the sky, the point about which all stars rotate, the north celestial pole, conveniently located close to a bright star, Polaris, at this point in history.

The usefulness of the second relationship comes in navigation, since a measurement of this angle of this easily found position above the horizon provides one's latitude. (The other coordinate that fixes one's position on the Earth, longitude, proved to be much more troublesome in the history of exploration. Accurate measurements of longitude were only possible after the development of accurate and portable chronometers or clocks. The modern Global Positioning System, GPS, depends on very precise measurements of distances from the GPS receiver to several satellites that all carry very accurate clocks.)

The relationships that you explored in this project can be represented geometrically by the use of the diagrams below.

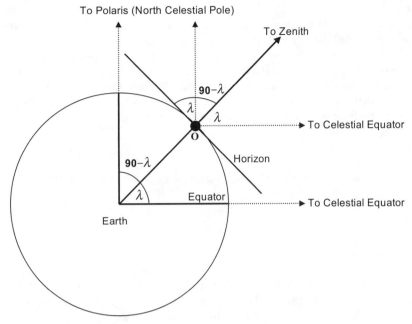

Figure 2. Altitude of Polaris and Latitude of Northern Hemisphere Observer

Figure 2 shows a section through the center of the Earth along a meridian of longitude. A northern hemisphere observer is at the point marked O. By definition, a line drawn from the center of the Earth through the observer points directly overhead to the zenith. Latitude, λ, is the angle between the equator and this line. To help you see this, the coordinate system defined by the reference plane of Earth's equator has been translated to the position of the observer at O in the figure. At O, one can see that, if the angle between zenith and celestial equator is λ as shown in Figure 1, then the angle between zenith and north celestial pole is equal to $(90° - \lambda)$. In turn, since the angle between the zenith and the northern horizon is 90°, the angle between this horizon and the north celestial pole is also simply λ.

Question 36. Suppose you knew that a particular star was on the celestial equator. If you measure its altitude when it reaches its highest point above the horizon and find it to be 32°, what is your latitude if a) the star reaches its highest altitude in the southern sky? b) the star reaches its highest altitude in the northern sky?

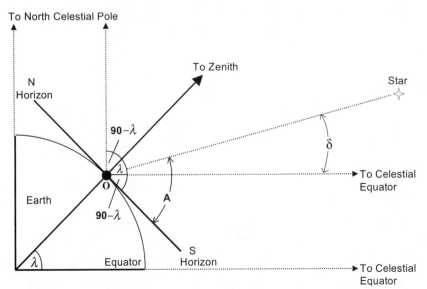

Figure 3. Relationship between Altitude of Star and Latitude of Observer

The relationship between the altitude, A, of a star above an observer's horizon and the latitude λ of the observer's position on Earth involves an extra angle, the declination, δ, of the star, its angle above or below the celestial equator.

On the central cross-section through the quadrant of the Earth shown in Figure 3, represented with the north celestial pole vertical and the celestial equator horizontal, the altitude, A, of a star above the southern horizon for a northern hemisphere observer at O is shown. The angle between the celestial equator and the zenith is again λ, the observer's latitude, as shown. Considering the 90° angle between zenith and the southern horizon, the angle between the celestial equator and the southern horizon is $(90° - \lambda)$. As can be seen from the diagram, the angle between the celestial equator and the southern horizon is also equal to angle $(A - \delta)$, where δ is the declination north or south of the celestial equator. Therefore,

$$(90° - \lambda) = (A - \delta)$$

where δ is positive for a star north of the celestial equator or negative if it is south of the celestial equator. Rearranging this equation gives

$$A = (90° + \delta) - \lambda,$$

from which you can see that as λ increases, A decreases.

For the north celestial pole, $\delta = 90°$. The angle between the southern horizon and the north celestial pole is $180° - \lambda$. The angle between the northern horizon and the north celestial pole is therefore $180° - (180° - \lambda) = \lambda$, in agreement with the argument in the previous section.

Another viewpoint for this equation is that a star of declination δ will just appear upon the horizon with $A = 0$ when it is on the meridian for an observer at a latitude of

$$\lambda = (90° + \delta)$$

Question 37. A star with declination of –32° (32° south) is observed to rise just above the southern horizon at midnight from a particular location. What is the latitude of this location?

Question 38. You are at a location on Earth at latitude 58° S. At midnight, a star rises just barely above the northern horizon. What is the declination of this star?

Diurnal Motion 5

Every day, the Sun rises in the east and moves (or appears to move) toward the west across the sky, finally setting below the western horizon. All objects in our sky—the Moon, planets, stars and galaxies—share this **diurnal motion,** but we notice this motion only at night for most objects other than the Sun.

The ancient Greeks explained diurnal motion by assuming that the Earth was stationary and that the rest of the universe rotated around it. We now know that diurnal motion is caused by the

Section 2–3 and Figure 2–4 of Freedman, Geller, and Kaufmann, *Universe*, 9th Ed., discuss diurnal motion.

Earth's rotation on its axis, carrying observers eastward. If you are unsure of this, a simple experiment may convince you that it is true. Try turning yourself slowly toward the left while looking straight ahead and watch your surroundings. Every object that starts on your left will appear to move slowly toward your right as you turn.

Thus, everything you see appears to move in the opposite direction to your motion. The reason that we see the Sun rise in the east each day and move slowly westward until it sets is because the Earth is turning toward the east.

The appearance of the Sun in our sky every day is fundamental to life on Earth. We define our time system by the time interval between successive appearances of the Sun, i.e., the rotational period of the Earth with respect to the Sun, known as the **solar day.** However, the Earth's motion in its orbit

Section 2–7 of Freedman, Geller, and Kaufmann, *Universe*, 9th Ed., discusses solar time and Box 2–2 discusses sidereal time.

around the Sun means that the Sun, as seen from Earth, appears to move against the background stars. Thus, the solar day is not the intrinsic rotation period of the Earth since it uses the Sun, a moving object, as a marker. The more fundamental rotation period of the Earth with respect to the background stars is known as the **sidereal day.**

In this project, you will observe the diurnal motion of the Sun and other celestial objects. You will also measure the sidereal day and compare it to the solar day.

A. Characteristics of the Diurnal Motion of Objects in the Sky

Observations of the diurnal motion of objects in the sky can be used to evaluate some of the characteristics of the Earth's rotation.

1. Launch *Starry Night Enthusiast*™.
2. Select **View > Solar System > Satellites** to remove artificial Earth-orbiting satellites from the view.
3. Set the **Date** to **September 21** and the **Time Flow Rate** to **300×**.
4. Set the **Gaze** to the East and click the **Sunrise** button in the toolbar.
5. Select **Labels > Planets-Moons**.

6. Observe the motion of the Sun for a while after sunrise and then increase the **Time Flow Rate** to 3000× to watch the Sun's motion across the whole sky. Use the **Gaze** buttons to keep the Sun in the view.

7. **Stop Time** once the sun sets below the horizon.

Question 1. In which direction does the Sun move?

Question 2. If the diurnal motion of the Sun is caused by the rotation of the Earth, in which direction is an observer on the Earth moving?

Question 3. Does the Sun appear to move at a steady rate or does it appear to speed up, slow down, or move irregularly?

Question 4. Is there any way to tell from your observations whether this motion is the result of the Sun moving across the sky or the Earth rotating beneath the sky?

The Moon and several of the planets may appear as sunset approaches, and after sunset, stars become visible. It is instructive to examine the motion of these objects briefly to show how they apparently share a common motion produced by the Earth's rotation.

8. Set the **Gaze** to the East.

9. **Run time forward at 300×.** Observe the motion of the objects in the sky. Use the **Gaze** buttons and the **hand tool** to look all around the sky.

10. When daylight returns, use the keyboard shortcut **Ctrl-D** to **Hide Daylight.**

11. Set the **Gaze** to the South and observe the motion of objects in this region of the sky.

Question 5. In which direction do the stars appear to move?

Question 6. In which direction do the Moon and planets appear to move?

To the Earthbound observer, the natural heavenly bodies appear to share a common motion that results from the Earth's rotation, at least over this limited time frame. However, many of these objects move with respect to the background. In the next sequence, you will observe the paths of these different natural celestial bodies across the sky. Your observations will help you to determine which of the celestial objects, or class of objects, most clearly reflects the rotation of the Earth on its axis by their diurnal motion. You can first examine the motion of the Sun in more detail by using *Starry Night Enthusiast*™ to trace its motion across the sky.

12. Select **Favourites > Observing Projects > Diurnal Motion > Sun's Diurnal Path** from the menu.

The view shows the Sun in the sky looking south from Reliance, Canada, at latitude 62° N, on September 27, 2008, at 12:00:00 PM local standard time.

13. **Run time forward** at **15,000×** until the Sun returns to near its original position. *Starry Night Enthusiast* ™ traces the path of the Sun through time with a green line. **Stop Time** once the Sun has traced a complete path around the sky.

14. Turn off the **Local Horizon** in the **Options** pane. Use the **Hand Tool** to explore the path that the Sun has made across the sky.

Question 7. What is the general shape of the path that the Sun traces in the sky?

Question 8. Does the Sun return to exactly the same position in the sky day after day?

15. Select **Favourites > Observing Projects > Diurnal Motion > Diurnal Paths of Moon and Planets** from the menu.

The view is the same as that in the previous sequence except that the Moon, Mercury, Venus, Mars, and Saturn are labeled in addition to the Sun. Note that the time is midday, 12:00:00 PM.

16. **Run time forward** at 20,000× and **Stop Time Flow** when the Sun is close to its original position. Adjust the minutes and seconds of the **Time** in the toolbar to place the Sun as close as possible to its original position. You might want to **Zoom in** on the Sun to adjust the time more precisely to the moment when the center of the Sun's disc reaches the starting point of its path. Then **Zoom out** to the full view again. You may want to remove daylight by typing **Ctrl-D** to make this observation easier.

17. Use the **Hand Tool** to explore the paths that these objects have traced out in the sky. Turn off the **Local Horizon** in the **Options** pane to see the parts of the paths that are obscured by the horizon.

18. Select **File > Revert** and press the **D** key on the keyboard to step forward in time by 1 day to display the positions of these objects after this interval.

Question 9. How long does it take for the Sun to return to its starting point, in solar time?

Question 10. What is the overall shape of the path that the planets trace in the sky?

Question 11. Do the planets and the Moon all return to approximately their original positions at the same time as the Sun? Why do you think this is so?

Question 12. Which of the labeled objects seems to be farthest from its starting position after this time interval? (Step 18 will help to answer this question.)

Question 13. Based on their observed deviation from their original positions after one day, which of these objects most accurately reflects the rotation of the Earth?

In the following sequence, you will observe the diurnal motion of the background stars and explore the definition of sidereal time and the length of the sidereal day, the time that represents the intrinsic rotation of the Earth.

19. Select **Favourites > Observing Projects > Diurnal Motion > Diurnal Paths of Stars.**

The view is the same as that of the previous sequences, but now the time has been set to 5:00:00 PM. *Starry Night Enthusiast* ™ has been set to hide the Moon, the Sun, and planets and to show *only* the brightest stars in the sky.

20. **Run time forward** at 1000× and watch the stars as they move across the sky. After a short time, the stars should begin to leave a track of their motion. (Note: It is possible that the graphics equipment of your computer will not support the display of these trails. If this is the case on your computer, jump to the alternate instructions beginning at Step 27.)

21. Select **File > Revert** from the menu to return to the initial configuration and erase the star trails from the screen.

22. Select **View > Zenith/Nadir** from the menu.

23. Set the **Gaze** first to the North and then to the Zenith so that the **Gaze** panel shows **Alt: 90° Az: 0°**.

24. Select **View > Hide Daylight** or use **Ctrl-D** to remove daylight.

25. **Run time forward** at 1000× and observe the paths traced out by the diurnal motion of the stars in this region of the sky. **Stop Time** once the stars have traced out a complete path.

26. If your graphics equipment successfully showed star trails, skip the next sequence of instructions in steps 27–31.

27. Select **Favourites > Observing Projects > Diurnal Motion > Diurnal Paths of Stars – No Trails**.

28. Remove daylight by typing **Ctrl-D**. The star **Sabik** just above the **S** point of the horizon is labeled in the view.

29. **Run time forward** until the stars have completed a full path around the sky and the star **Sabik** is again above the **S** point of the horizon. Adjust the minutes and seconds of the time to return **Sabik** to its original position above the south point of the horizon.

30. Move the mouse cursor over any star near the horizon and right-click to open the **Object Contextual Menu** for the star and **Select** this star to label it for future reference.

31. **Run time forward** at 1000× and observe the paths traced out by the diurnal motion of the stars in this region of the sky. **Stop time** once the stars have traced out a complete path.

Question 14. Is the motion of the stars smooth and regular, or do they appear to speed up and slow down as they move across the sky?

Question 15. What is the shape of the path that the stars trace out in the sky?

Question 16. Do the stars move in a clockwise or anticlockwise direction in the view?

Question 17. Is the path of the stars in this region of the sky a closed curve? In other words, do the stars return to exactly the same position in the sky after a period of time?

Question 18. Do all of the stars visible in the view return to their starting points simultaneously?

Question 19. From your observations of the diurnal motions of the Sun, the Moon, several planets, and the stars, which do you think most clearly reflect the actual rotation of the Earth on its axis?

Question 20. Does the Earth rotate smoothly on its axis or does it speed up and slow down over the course of a day and a night?

B. Orientation in Space of the Rotation Axis of the Earth

The paths that the stars trace across the sky as a consequence of the rotation of the Earth are called **diurnal circles**. The center of these diurnal circles indicates the orientation of the Earth's axis in space. This axis passes through the north and south poles of the Earth and its extension intersects the celestial sphere at the **north** and **south celestial poles**.

Figure 2–9 of Freedman, Geller, and Kaufmann, *Universe*, 9th Ed., illustrates the north and south celestial poles and the celestial equator.

The plane that passes through the center of the Earth and oriented perpendicular to its rotational axis intersects the surface of the Earth at the equator. The extension of this plane intersects the celestial sphere at the **celestial equator**.

32. Select **Observing Projects > Diurnal Motion > Rotating Earth** from the **Favourites** pane. If the Earth does not appear at the center of the view, click in the main view window and then select **File > Revert**.

The view is from a point in space hovering 11,000 kilometers above the surface of the Earth. The red line on the image of the Earth indicates the Earth's equator, while the red line crossing the sky behind the Earth is the celestial equator. Blue and yellow pole sticks mark the north and south poles of the Earth, respectively, and indicate the axis on which the Earth rotates. The boundaries and labels of the constellations are displayed on the stellar background. From the perspective of a point on the equatorial plane of the Earth, you can see that the celestial equator and the Earth's equator are aligned.

33. Position the cursor over the image of the Earth. The cursor changes to the **Location Scroller.** You can hold down the mouse button and move the mouse to change your perspective and look all around the view. The Earth and the sky appear to rotate around the center of the Earth as the **Location Scroller** changes your viewing location around a sphere 11,000 kilometers above and concentric with the surface of the Earth. As you look around the view, you will notice that the north and south celestial poles are marked on the sky. You may need to use the **Hand Tool,** with the cursor moved off the Earth, to reposition the Earth in the view and then adjust the view with the **Location Scroller,** with the cursor on the Earth again, in order to see different parts of the sky.

Question 21. In which constellation in the sky is the north celestial pole?

Question 22. In which constellation in the sky is the south celestial pole?

Question 23. Toward which constellations is the rotation axis of the Earth pointed?

34. Select **File > Revert** and then **Run time forward** to observe the Earth's rotation against the background stars. The stars remain stationary in the view because the observing location is fixed with respect to the stars as time advances.

35. Now use the **Location Scroller** and **Hand Tool** to observe the view of the Earth rotating in space from different perspectives, particularly from positions looking down on the Earth from above each of its poles.

Question 24. When looking nearly along the rotational axis of the Earth toward the north celestial pole from a position over the south pole, a) in which sense does the Earth rotate, clockwise or anticlockwise? and b) toward which compass direction does an observer on the Earth move?

Question 25. When looking nearly along the rotational axis of the Earth toward the south celestial pole from a position over the north pole, the direction approximately opposite to the view described in the previous question, a) in which sense does the Earth rotate, clockwise or anticlockwise? and b) toward which compass direction does an observer on the Earth move?

Question 26. Use the following sequence (Steps 36–41) to help you answer these questions:

a) Where on the surface of the Earth would you expect to see the diurnal circles traced out by the stars centered on the zenith? [Hint: Where on the Earth would the zenith coincide with a celestial pole?]

b) In which sense would the stars appear to rotate around the zenith from this location or locations, clockwise or anticlockwise?

c) At this location or locations on the Earth, how are the diurnal circles of the stars oriented relative to the horizon?

36. **Stop time** and use **Options > Viewing Location...** to change your location to **the surface of** the **Earth** and select a location from the **List** at which you suspect that the conditions specified in the previous question prevail.

37. Select **View > Zenith/Nadir** to label this location in your sky and select **View > Celestial Poles** to turn this option **off.**

38. Set the **Gaze** to the Zenith.

39. **Run time forward** to check your answer to parts a) and b) of the previous question.

40. Select **View > Celestial Poles** to turn this option back on.

41. To answer part c) of the previous question, change the **Gaze** to look around the cardinal points of the horizon, North, East, South, and **West** and observe the diurnal motion of the stars near the horizon. Slow the **Time Flow Rate** if necessary in order to see this motion.

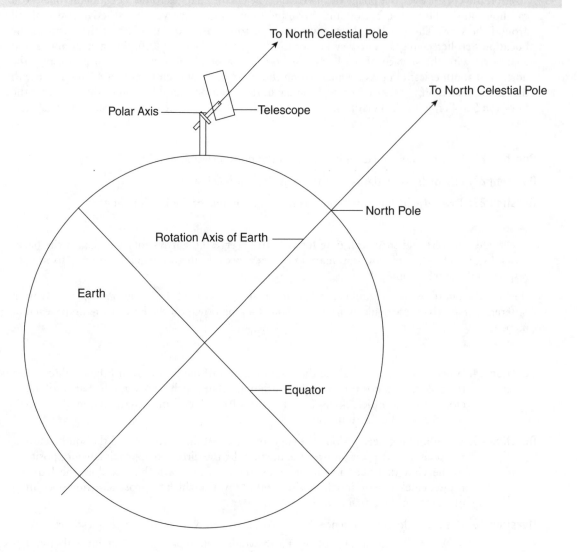

Figure 1. Polar Axis of a Telescope

The orientation of the rotation axis of the Earth is important in one particular application, the mounting of an astronomical telescope. In an **equatorial mount,** the **polar axis** of the telescope is aligned accurately with the Earth's axis, as shown schematically in Figure 1. An observer in the northern hemisphere would point this axis of the telescope to the north celestial pole and a southern hemisphere observer would point this axis to the south celestial pole. Rotation of the telescope about this polar axis at the appropriate rate allows it to follow the diurnal path of objects in the sky by counteracting the rotation of the Earth. Without such tracking, the rotation of the Earth quickly carries objects out of the small field of view of a telescope. The second axis of rotation on the telescope mount, the **declination axis,** is aligned perpendicular to the polar axis, allowing the telescope to be pointed to objects anywhere in the sky.

Question 27. An observer in the northern hemisphere uses a telescope with a clock drive on its polar axis to observe the sky. Could this observer use the same telescope as easily in the southern hemisphere? Why or why not?

C. Diurnal Motion and the Duration of the Day

In this section, you will determine how long it takes for the Earth to make one complete rotation on its axis. This period is defined as the **day**. Each day is divided into 24 equal hours. Each hour, in turn, is divided into 60 equal minutes and each minute is divided into 60 equal seconds.

> Section 2–7 of Freedman, Geller, and Kaufmann, *Universe*, 9th Ed., discusses solar time and Box 2–2 discusses sidereal time.

The standard time defined for everyday purposes, **solar time**, is related to the Earth's rotation with respect to the Sun. A second time system, **sidereal time**, is related to rotation of the Earth with respect to the stars and is used mainly in astronomy.

Measurement of the length of the day requires a reference plane fixed with respect to an observer on Earth. The day is then the time between successive passages of the reference plane past the reference object. A convenient plane is the plane that contains the north and south celestial poles and the observer's zenith. This great circle is known as the **meridian**.

> The meridian is illustrated in Figure 2–21 of Freedman, Geller, and Kaufmann, *Universe*, 9th Ed.

The fact that this plane passes through the observer's zenith fixes the meridian to the observer's location on the Earth. Observers on the surface at different longitudes will have different meridians, while observers along precisely the same longitude on the Earth will share a meridian. As the Earth rotates, the meridian sweeps across the sky. For an observer on the surface, however, the meridian appears fixed and it is the sky that appears to rotate past the meridian.

> Meridian transits are discussed in Section 2–7 of Freedman, Geller, and Kaufmann, *Universe*, 9th Ed.

When an object in the sky crosses an observer's meridian, it is said to **transit**. Therefore, we can use the duration between transits of a celestial object to provide a measure of the rotational period of the Earth and, therefore, the length of the day. As you will see, the duration that we call one day depends on the reference object chosen.

The **solar day** is defined as the time between successive transits of the Sun. Unfortunately, the Sun is not a reliable timekeeper. The motion of the Earth in an elliptical orbit inclined to the celestial equator causes the

> The types of solar day and the timekeeping reliability of the Sun are discussed in Section 2–7 of Freedman, Geller, and Kaufmann, *Universe*, 9th Ed.

Sun to appear to move non-uniformly through the year when viewed from Earth. For accurate timekeeping, a mean rate of rotation of the Earth with respect to the Sun is defined by international agreement and the accepted average time period for the length of the day is known as the **mean solar day**.

> The sidereal day is defined in Box 2–2 of Freedman, Geller, and Kaufmann, *Universe*, 9th Ed.

The fact that the Sun moves with respect to the background stars means that the duration between successive transits of a star is different from that for the Sun. The period between successive transits of a chosen star, the **sidereal day**, does not vary though the year. Apart from small effects such as the motion of precession of the Earth as it spins, this time represents the intrinsic rotation rate of the Earth. The following sequence demonstrates the motion of the Sun with respect to the star background.

42. Select **Favourites > Observing Projects > Diurnal Motion > Transits of Sun and Aldebaran** from the menu.

The view looks south from Calgary, Canada, on May 31, 2004, at 12:33:45 PM local standard time. The center of the Sun is on the meridian. With daylight removed from the view, you can see the background stars as well as the Sun. The time has been chosen to place the Sun on the meridian, while the date has been chosen so that the star Aldebaran is on the meridian at this time. In the next step, you will compare the rotational period of the Earth as measured by the Sun with the rotational period measured with respect to Aldebaran.

43. Start **Time Flow** at the set rate of 3000× and allow time to flow until the Sun approaches the meridian again. **Stop Time** when the Sun gets close to the meridian and adjust the minutes and seconds of **Time** in the toolbar to move the center of the Sun onto the meridian.

44. Use the **Angular Measurement Tool** to measure the distance between the star Aldebaran and the meridian at the point in time when the center of the Sun has returned to transit the meridian. [Hint: You may want to **Zoom in** on Aldebaran to make this measurement.]

Question 28. In the time required for the Sun to move completely around the sky and return to the meridian, what has happened to the star Aldebaran?

Question 29. At the time of the second transit of the Sun, what is the angular distance between the star Aldebaran and the meridian?

Question 30. Assuming that the rotational period of the Earth with respect to Aldebaran is representative of the period that would be measured using any other star, which of the following conclusions are correct?

   a) The Earth's diurnal period with respect to the Sun (the solar day) is longer than its diurnal period with respect to the stars (the sidereal day).

   b) The Sun remains stationary against the background stars during the interval of one rotational period of the Earth.

   c) The rotational period of the Earth as measured by the stars is shorter than the rotational period of the Earth as measured by the Sun.

   d) A solar day is longer than an Aldebaran day.

Question 31. The Earth makes one complete orbit of 360° around the Sun in the interval of one year, which is equal to approximately 365.25 solar days.

   a) How far around the orbit does the Earth move in the interval of one solar day?

   b) How does this compare with the measurement you obtained for the distance of Aldebaran from the meridian at the time of the second transit of the Sun?

The next sequences will allow you to quantify the difference between the duration of the Earth's rotational period as measured with respect to the stars and the duration of this period as measured with respect to the Sun. To measure the time of this duration, you will use a standard clock running at the solar rate displayed in the Time and Date panel of *Starry Night Enthusiast*™. You will use this standard measure of time to measure the duration of a day as defined by the Sun and a sidereal day as measured by the star Menkar.

45. Select **Favourites > Observing Projects > Diurnal Motion > Apparent Solar Day** from the menu.

The view looks south from Calgary, Canada, at 12:19:32 PM standard time on November 3, 2007. The Sun is on the meridian.

46. **Step time forward** by 24 hours and verify that the Sun returns very close to the meridian in this period of 1 mean solar day. (The fact that the Sun does not return *exactly* to the meridian is evidence of the difference between a "real," or apparent, solar day and the mean solar day. This difference is discussed in a separate chapter on the Analemma.)

Question 32. At this time of the year, is the apparent solar day, as defined by a return of the Sun to the meridian, close to the mean solar day defined by 24 hours of solar time?

In the next sequence, you will measure the duration of the day based on the diurnal motion of the star Menkar using the standard clock in the *Starry Night Enthusiast* ™ toolbar.

47. Select **Favourites > Observing Projects > Diurnal Motion > Transits of Menkar** from the menu.

The view looks south from Calgary, Canada, at 12:49:51AM standard time on November 3, 2007. The star Menkar is labeled and on the meridian at this precise time on this day. The star Tau3 Eridani is also labeled and on the meridian.

48. Record the time of this first transit of Menkar in Data Table 1 below.

49. **Run time forward** at 3000× its normal rate until Menkar approaches the meridian. Use the minutes and seconds in the **Time and Date** pane in the toolbar to advance time until Menkar is precisely on the meridian. Record the time of this second transit in Data Table 1. [Hint: You may want to **Zoom in** on the view to observe the transit accurately.] Note also the position of Tau3 Eridani with respect to the meridian.

50. Use your observations to calculate the duration of a sidereal day, based on the diurnal motion of the star Menkar.

Data Table 1: Time Between Transits of the Star Menkar and the Length of the Day based on the Diurnal Motion of the Stars

Time of Second Transit	
Time of Second Transit + 24 hours	
Time of First Transit	
Difference (Duration of Day)	

Question 33. What is the length of the day based on the diurnal motion of the star Menkar?

Question 34. Did the star Tau3 Eridani show evidence of a different duration between transits from that of Menkar?

Question 35. How does the length of the day as measured by the Sun compare with the length of the day as measured by the stars?

Question 36. If an astronomer was interested in tracking a star with a telescope, at what rate would the telescope drive need to turn in degrees per hour of time as measured by the standard clock in the *Starry Night Enthusiast* ™ interface?

Question 37. If a solar astronomer wished to track the Sun with an appropriately filtered telescope, at what rate would the clock drive of the polar axis need to turn in degrees per hour of time as measured by the *Starry Night Enthusiast* ™ interface?

Question 38. From your observations, is the diurnal motion of the Sun or that of the stars a more accurate and reliable measure of the actual rotational period of the Earth on its axis?

D. Conclusion

In this project, you have explored the diurnal motion of celestial objects. From your observations, you have been able to deduce the orientation of the rotation axis of the Earth in space. You have seen how this rotation of the Earth affects the design of mounts and tracking mechanisms of telescopes. Finally, you compared the rate of the diurnal motion of the Sun with that of the stars and used these data to determine the intrinsic rotation period of the Earth.

Earth's Orbital Motion 6

Our view of the night sky is constantly changing. Two of the most evident changes are the apparent motion of the stars and constellations toward the west over the course of a single night and the much slower shift of stars and constellations toward the west over the course of a year when viewed at the same time each night. The first of these changes is caused by the Earth's rotation. This makes the sky appear to rotate around the Earth's polar axis once every day at a rate of 15° per hour and makes the Sun, Moon, and stars rise in the east and set in the west each day. The second change is caused by the Earth's orbital motion around the Sun. Since the Earth makes a full orbit of 360° in about 365 days, our view of the universe at any given time of the night changes by about 1° per day. The combined result of these two changes is that, in terms of our Sun-based time, each star rises about four minutes earlier each night.

> Section 2–3 and Figure 2–5 of Freedman, Geller, and Kaufmann, *Universe*, 9th Ed., discuss the Earth's orbital motion.

A. The Shifting Constellations

If you go out at the same time each night and observe the sky, you will see that the positions of the constellations change as the months pass.

1. Launch *Starry Night Enthusiast*™.
2. Set the **Time Flow Rate** to **1 day.**
3. Set the **Time** in the toolbar to **12:00:00 AM.**
4. Set the **Gaze** direction to the South if your home location is in the northern hemisphere and to the North if you live in the southern hemisphere.
5. Select **View > Constellations > Astronomical** and **View > Constellations > Labels.**
6. **Step time forward** one day at a time.

Question 1. Toward which direction do the constellations appear to move, night by night?

7. Select **View > Local Meridian.**
8. Note the **Date** from the display in the toolbar and select a constellation on or near the meridian.
9. **Run time forward** and **Stop Time** when the selected constellation returns to the meridian. Compare the **Date** shown in the toolbar with that of the previous step.

Question 2. Approximately how long does it take for a particular constellation to return to the meridian?

This daily shift of the background stars is the result of Earth's orbital motion around the Sun. The interval between frames in the previous animation was one solar day, the time required for the Earth to make one complete rotation on its axis with respect to the Sun. Consequently, since the time is midnight, each frame of the animation shows the sky along a line of sight directly away from the Sun. As the Earth orbits the Sun, the line of sight sweeps out a complete circle around the celestial sphere. The next sequence shows this from a position in space directly between the Earth and the Sun, looking out toward the midnight sky of Earth.

10. Select **Favourites > Observing Projects > Earth's Orbital Motion > Earth's Motion.**

This view is from a position 15,000 kilometers above the prime meridian of Earth (illustrated by the vertical red line on the image of the Earth). The date is December 20, 2006, at noon over this location on Earth. The Sun is directly behind the observing location. From this point of view, the star Betelgeuse, in the constellation Orion, is on the meridian. You are looking down on the sunlit half of the Earth at midday but you can see the sky beyond the Earth. For a person on the opposite side of the Earth, it is midnight and Betelgeuse is transiting the meridian.

11. **Step time forward** by **1 day.**

In the time required for the Earth to rotate once on its axis with respect to the Sun, it has also moved a short distance along its orbit. Because the view is locked on the Earth and you are traveling through space along with it, this motion shows up as a movement of the background in the opposite direction.

12. **Run time forward** and **Stop Time** when the Earth returns to a position where its meridian aligns once again with Betelgeuse and note the **Date** displayed in the toolbar.

With the view locked on the Earth, the previous animation illustrates how the shift of constellations arises from the Earth's orbital motion.

Question 3. How long does it take for Betelgeuse to return to alignment with the meridian?

Of course, the distant stars are a fixed background, and it is the Earth that is moving.

13. Select **File > Revert.**
14. Position the cursor over Betelgeuse and **right-click** (**Ctrl-click** on a Macintosh) to open the **Object Contextual Menu**, and select **Centre.**
15. **Step time forward** several days.

The view is now locked onto the background stars. The viewing location remains fixed at a point 15,000 kilometers from the surface of the Earth on a line connecting the Sun and the Earth. Now, however, the Earth will appear to move, while the background stars remain fixed in the view. Your viewpoint is still moving with the Earth but your field of view is always centered on Betelgeuse. You will move to the other side of the Sun from Betelguese as time advances.

16. **Run time forward** and **Stop Time** when the local meridian and the Sun (labeled in the view) align with Betelgeuse.

Question 4. On what date does the Sun align with Betelgeuse on the local meridian?

17. **Run time forward** until Earth reappears in the view and aligns once more with Betelgeuse. Note the **Date** shown in the toolbar.

Question 5. How long did it take for the Earth to move completely around its orbit to return to alignment with Betelguese?

B. The Solar and Sidereal Day

In the previous animations, the time interval between frames was 1 solar day, the time required for the Earth to rotate completely on its axis with respect to the Sun. We can change this to make the time interval 1 sidereal day, the time required for the Earth to rotate completely with respect to the background stars.

18. Select **File > Revert**.
19. Change the **Time Flow Rate** to 1 sidereal day.
20. **Step time forward** by 1 sidereal day.

After one full rotation of the Earth, this time step of 1 sidereal day brings the prime meridian of Earth back into alignment with the background stars.

Question 6. What is the time that is displayed in the toolbar?

Question 7. What is the difference between the initial time of 12:00:00 PM and the time displayed in the toolbar?

Question 8. The Earth rotates in the same direction as its orbital motion, that is, anti-clockwise when looking down on the north pole. If the Earth rotated in the opposite direction to its orbital motion, would 1 sidereal day be shorter or longer than 1 solar day? (Hint: Draw a diagram to help you answer this question.)

21. **Step time forward** several more times.

You may not notice any changes in the view.

22. **Run time forward.**

As time runs forward in intervals of 1 sidereal day, the Earth's orientation with respect to the background stars remains fixed. From this viewing location, it is the Sun that appears to move. The sunlit hemisphere of the Earth indicates its direction.

23. Select **Favourites > Observing Projects > Earth's Orbital Motion > Sidereal and Solar.**

The view is similar to that of the previous sequence except that now the viewing location is from a position *hovering* above the surface of the Earth, so that the orientation of the viewing location to the center of the Earth remains fixed relative to the background stars. The Earth will thus rotate under this observing location.

24. Open the **Status** pane and expand the **Location** layer and note the latitude and longitude over which the viewing location is hovering.
25. **Step time forward** by 1 sidereal day.

Notice that the longitude of the **Location** in the **Status** pane does not change in the period of 1 sidereal day. This is because the Earth has made one complete rotation with respect to the stars. From your hovering position, the same longitude has been rotated under your viewing location, which is in a fixed direction from the center of the Earth with respect to the stars. In this time interval, the Earth will have moved along its orbit, so the Sun will have moved slightly from its initial position directly behind the viewing location. This change will become more apparent over longer times.

26. **Run time forward** for a few months and note the change in the orientation of the sunlit hemisphere on the Earth.
27. Select **File > Revert** to return to the initial set-up.
28. Change the **Time Flow Rate** to **1 day.**
29. **Step time forward.**

Time now advances by 1 solar day. In this interval, the Earth will have moved along its orbit a short distance. The orientation of the line joining the Earth and the Sun has shifted in space with respect to the background stars. For the Earth's prime meridian to point directly at the Sun at noon, the Earth must rotate a small amount to compensate for its orbital motion. Notice that the prime meridian, which is pointing toward the Sun, has moved. Also notice from the **Location** layer of the **Status** pane that the position on the Earth over which the observing location is hovering has also changed. The difference in longitude of the location indicates the amount of extra rotation that the Earth must complete to compensate for its orbital motion and return the Sun to the meridian at noon.

> Box 2–2 in Freedman, Geller, and Kaufmann, *Universe*, 9th Ed., discusses the sidereal time and the difference between this and solar time.

Question 9. Toward which compass direction on the Earth has the prime meridian of the Earth shifted?

Question 10. How much further did the Earth rotate in order compensate for the orbital motion of the Earth?

You can now measure the angular distance moved by the Earth in 1 day by using a background star as a reference point against which to measure this motion.

30. Select **Favourites > Observing Projects > Earth's Orbital Motion > Earth's Motion.**
31. Use the **Object Contextual Menu** over **Betelgeuse** to **Centre** the view on this star. Then **Zoom In** to a field of $2° \times 2°$.

32. Use the **Angular Measurement Tool** to measure the distance between Betelgeuse and the local meridian (the grey vertical line).

33. **Step time forward** by 1 solar day and repeat the previous measurement. Since the meridian is fixed with respect to the Earth and on the same side of Betelgeuse in both frames, you can calculate the angular distance that the Earth has moved in its orbit in the duration of 1 day by subtracting the first measurement from the second, thereby determining the motion of the Earth in 1 day.

Question 11. How far along its orbit did the Earth move in 1 solar day?

Question 12. Given the potential inaccuracy of the measurements you made, how do your answers to the previous two questions compare?

Question 13. Since the Earth rotates on its axis a full 360° with respect to the Sun in 24 hours of solar time, how long does it take to rotate the extra angle that you noted in Question 10?

Question 14. How does your answer to the previous question compare to the difference between the length of the sidereal and solar days?

Question 15. Calculate the ratio between the length of a solar day and a sidereal day. Multiply this ratio by the number of solar days in 1 year (365.25) to calculate the number of sidereal days in 1 year. How do you account for this number?

34. Select **File > Revert**.

35. Open the **Find** pane and click the menu button to the left of the listing for the **Sun** and select **Magnify**.

Now you are looking toward the Sun, which is in the direction of the labeled star, TYC6841-743-1. In 1 year, the Earth will have moved completely around the Sun in its orbit so that the Sun will be in the same direction against the background stars as it was a year earlier. You can first step time forward by 1 solar year, that is, 365.25 days, to verify that the Sun returns to the same position when viewed from Earth.

36. Set the **Time Flow Rate** to 365 days and **Step time forward**. Then change the **Time Flow Rate** to 6 hours and **Step time forward**.

You can now advance time by 1 sidereal year, that is, the number of days calculated in Question 15, and check that the Sun returns to the same position in our sky after this time.

37. Change the **Time** in the toolbar to **12:00:00** PM and the **Date** to **December 20, 2006**.

38. **Step time forward** by **6 hours** to account for the quarter day part of the year and then change the **Time Flow Rate** to **sidereal days** and enter the number of sidereal days in 1 year that you calculated in Question 15 into the numerical field of the **Time Flow Rate** panel. **Step time forward** to check your answer. The Sun should once again be in the same direction in the sky as the star TYC6841-743-1.

C. The Sun's Apparent Motion Across the Sky and Kepler's Second Law

If we could see the stars during the day, the effect of the Earth's orbital motion around the Sun would be an apparent motion of the Sun against the stars. In real life, the blue sky of daylight, which is the result of scattering of sunlight by the Earth's atmosphere, hides the background stars. *Starry Night Enthusiast™* allows you to remove the effect of this obscuring atmosphere.

---

39. Select **Favourites > Observing Projects > Earth's Orbital Motion > Sun's Apparent Motion.**

The view shows the sky looking south from Chicago at noon on December 20, 2006. Daylight has been turned off to reveal the background stars. The Sun is near the center of the view and labeled. The **Time Flow Rate** is set to 1 sidereal day. Thus, the background sky will return to exactly the same position in the view and the stars will appear to remain stationary as time advances. The motion of the Sun against these background stars is then much easier to see.

---

40. **Step time forward** 1 sidereal day at a time.

Question 16. In which direction does the Sun appear to move in our sky in 1 sidereal day?

You can measure the apparent speed of the Sun across the sky by choosing an appropriate reference star close to the Sun on its path. Since you are measuring this solar motion from a moving Earth, this motion simply reflects the speed of the Earth in its orbital path. This motion will carry the Earth around a full orbit in a time of 1 year. However, because the Earth's orbit is elliptical, this speed will vary through the year, following Kepler's second law of planetary motion. You can make a series of simple measurements of the Sun's apparent motion through the year to verify this law.

> Kepler's second law is discussed in Section 4–4 of Freedman, Geller, and Kaufmann, *Universe,* 9th Ed.

The consequence of this law is that the speed of a planet, v, in AU per sidereal day, is inversely proportional to its distance from the Sun, R, in AU. You will measure the Sun's angular speed across the sky, α, in arcseconds per sidereal day.

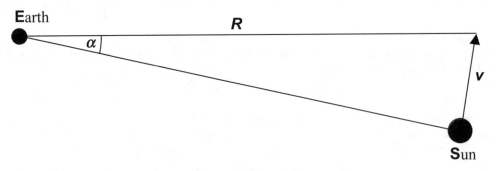

Figure 1. Angular Speed of Sun Across the Sky in 1 Sidereal Day

From Figure 1 above, it can be seen that this angular measurement, α, if measured in radians per sidereal day, using the small-angle relation, is given by

$$\alpha = \frac{v}{R}$$

Rearranging this equation gives

$$v = \alpha \times R$$

Since you will measure angular speed in arcseconds per sidereal day and 1 radian = 2.06×10^5 arcseconds,

$$v = \frac{\alpha \times R}{2.06 \times 10^5}$$

Kepler's Second Law can be represented by $v \times R = k$, where k is a constant. For the present measurements, this becomes

$$\frac{\alpha \times R^2}{2.06 \times 10^5} = k$$

which simplifies to

$$\alpha \times R^2 = k_1$$

where k_1 is another constant.

You can verify Kepler's second law by demonstrating this simple relationship with data measured through the year of the Sun's apparent motion across the sky. The following procedure will accomplish this.

41. Select **File > Revert** in the view named **Sun's Apparent Motion**.

42. Open the **Object Contextual Menu** over the **Sun** and select **Centre**.

43. **Zoom In** to a field of view of about **2° × 1°**.

44. Select **View > The Ecliptic** to display the path of the Sun's apparent motion across the sky. The ecliptic is the projection of the plane of the Earth's orbit onto the celestial sphere and contains the Sun.

45. Open the **Field of View (FOV)** pane and display the **1°** field of view.

46. Open the **Info** pane and check that the information refers to the **Sun**.

Sequence 1

47. Note the **Distance from Observer** of the Sun under the **Position in Space** layer of the **Info** pane and record this into Data Table 1 under the column Earth-Sun Distance, R (AU).

48. Select a suitable star that is on or near to the **Ecliptic** and within the 1° FOV indicator. Open the **Object Contextual Menu** for this star and click the **Select** option to label the star. Open the **Object Contextual Menu** over the star again and select **Centre** to move the reference star to the center of the field of view.

49. Use the **Hand Tool** to measure the angular separation between the Sun and the selected star, and enter this distance, in arcseconds, into Data Table 1 under the column Star-Sun Angular Separation Measurement #1. To convert the measurement to arcseconds, multiply the number of arcminutes in the measurement by 60 and add this product to the number of arcseconds in the measurement.

50. **Step Time** by **1 sidereal day** either forward or backward to move the Sun to the other side of the centered reference star.

51. Use the **Hand Tool** to measure the angular distance between the Sun and the reference star again and enter it, in arcseconds, into Data Table 1 under the column Star-Sun Angular Separation Measurement #2.

52. Open the **Object Contextual Menu** for the Sun and select **Centre** to center the Sun in the view once more.

53. Press the **M** key on the keyboard **twice** to advance the **Date** in the toolbar two calendar months and ensure that the **Time** remains at midday, **12:00:00 PM**.

54. Return to the beginning of **Sequence 1** and repeat the steps to make measurements through a full year.

The following calculations will allow you to verify Kepler's second law from your data.

55. Add the two angular measurements together and record the sum in arcseconds into Data Table 1 under the column Angular Motion of Sun, α in arcseconds per sidereal day.

56. Calculate the square of the Earth-Sun distance and multiply it by α, the angular motion of the sun. Enter this value of $A = \alpha \times R^2$ in Data Table 1.

57. Calculate the average, A_V, of all your values of $\alpha \times R^2$.

58. As a measure of the spread of these values from the average value, calculate the percentage difference between each value and the average value by taking the difference between each value and this average value, A_V, divide this difference by A_v and multiply by 100. You can examine these values of $100(A - A_V)/A_V$ to see how far they vary from the mean value.

If these values of the calculation of $\alpha \times R^2$ are constant over this 1-year series of measurements, then you have shown that the Earth obeys Kepler's second law.

Data Table 1. Sun's Apparent Motion in 1 Sidereal Day Throughout the Year

Date	Earth-Sun Distance, R (AU)	Star-Sun Angular Separation Measurement #1 (")	Star-Sun Angular Separation Measurement #2 (")	Angular Motion of Sun, α ("/Sidereal day)	$A = a \times R^2$	Variation in A (%)
				Average, A_V		

Question 17. What is the speed of the Sun across your sky in December, in arcseconds per sidereal day?

Question 18. What is the speed of the Sun across your sky in June, in arcseconds per sidereal day?

Question 19. Do your results verify Kepler's second law by showing that $A = \alpha \times R^2$ remains constant over the full year to within 1%, 0.5%, or 0.1%?

D. Motion of the Sky in Solar Time

In the previous sequences, you advanced the time in sidereal days in order to keep the stars fixed in the sky. This allowed you to watch the apparent motion of the Sun relative to the stars. It is interesting to examine the change in appearance of the sky as time changes in units of solar days rather than sidereal days. In this case, the stars will move across your sky while the Sun remains at approximately the same azimuth, in the present case almost due south. The Sun will move through a range of elevation angles because of the tilt of the Earth's spin axis to its orbital plane. This change of elevation angle through the year is important in defining the seasons on the Earth.

59. Click the **Home** button in the toolbar to go to your home location on Earth.

60. Set the **Time Flow Rate** to **1 day.**

61. Set the **Time** in the toolbar to **12:00:00 PM.**

62. Select **View > Hide Daylight.**

63. Open the **Find** pane and click the menu button to the left of the listing for the **Sun** and select **Centre** from the menu to lock the gaze onto the Sun.

64. **Run time forward.**

Day-by-day, the Earth advances along its orbit. With time flowing in intervals of 1 solar day and the gaze locked on the Sun, the Earth's orbital motion around the Sun causes the background stars to appear to move in the opposite direction. This is like standing on a merry-go-round looking inward with your gaze fixed on its center. As you move around in one direction, the stationary background scenery appears to move in the opposite direction.

You can watch the Sun as it moves along the ecliptic, the projection of the Earth's orbit onto the sky. Note that this line is labeled with time in months.

65. **Stop Time.**

66. Select **View > The Ecliptic.** A green line showing the ecliptic appears in the view.

67. **Step time forward** at 1-day intervals and compare the **Date** shown in the toolbar with the position of the Sun with respect to the demarcations on the ecliptic.

Question 20. What do the labels represent?

Question 21. What part of the month does the demarcation labeled with the name of the month correspond to?

E. The Zodiac

The ecliptic passes through a limited set of constellations known as the Zodiac.

68. Select **View > Constellations > Zodiac.**

69. Select **View > Constellations > Boundaries.**

70. Select **View > Constellations > Labels.**

Starry Night Enthusiast™ displays the boundaries, labels, and astronomical illustrations of the constellations of the traditional zodiac. The zodiac is a band of constellations through which most of the brightest planets and the Moon appear to move as they orbit close to the ecliptic plane. This region of our sky played an important role in early astrology because the positions of the planets were used to predict the future and these zodiacal constellations or "signs" are still used in modern astrology.

71. Use the **Run, Stop,** and **Step Time** controls in the toolbar to move through time at 1-day intervals and make observations that will allow you to answer the following questions.

Question 22. How many constellations comprise the traditional zodiac?

Question 23. What are the names of the constellations in the traditional zodiac?

Question 24. How many constellations does the ecliptic actually pass through?

Question 25. Which constellation(s), if any, are absent from the traditional list? [Hint: To see the names of all of the constellations, select **View > Constellations > Astronomical.**]

Question 26. Is the Sun in the constellation of your astrological sign on your birthday?

Question 27. How does this influence your confidence in the validity of your horoscope for today?

72. Exit *Starry Night Enthusiast* ™ without saving changes to the files.

F. Conclusions

This project has shown you the apparent motion of the Sun through our sky in a way that is impossible to demonstrate in the real world because of the presence of scattered sunlight in the daytime. Furthermore, you have been able to measure the variation in speed of the Sun across the sky caused by the variation of the speed of our observing platform, the Earth, as it moves in its elliptical orbit and verify Kepler's second law of planetary motion with these measurements.

Seasons

7

L ife at temperate latitudes on the Earth is strongly influenced by seasonal variations. The year is conveniently divided into the seasons of spring, summer, autumn, and winter to describe these variations. Summers are warm and winters are cool, while spring and autumn, or fall, are seasons of change. The cause of these variations is the tilt of the Earth's axis of rotation with respect to the plane of the Earth's orbit. Since the direction of this axis is maintained in space as the Earth moves around in its orbit, the Sun illuminates different parts of the Earth throughout the year.

> The seasons are discussed in Section 2–5 of Freedman, Geller, and Kaufmann, *Universe*, 9th Ed.

In June, the north pole is tilted toward the Sun, giving more direct sunlight and warmer temperatures to northern latitudes and less direct sunlight and cooler temperatures to southern latitudes. Thus, it is summer in the northern hemisphere and winter in the southern hemisphere.

In December, the Earth has moved to the opposite side of its orbit but has maintained the direction of its spin axis. Thus, the north pole is tilted away from the Sun and the northern hemisphere receives less direct sunlight, while the south pole is tilted toward the Sun and the southern hemisphere receives more direct sunlight. At this time, northern latitudes are cooler and the southern latitudes are warmer, leading to winter in the northern hemisphere and summer in the southern hemisphere.

Two other factors affect the heating and cooling of the Earth during these seasonal variations at mid-latitudes. The length of daylight changes from winter to summer, thereby changing the amount of overall heating of the Earth's surface. The second effect is the delay between the time of maximum heating, when the Sun is highest in the sky at a particular location, and the time of highest temperatures. This climate delay is caused by the fact that it takes time to heat up the Earth's surface and oceans. This delay varies somewhat depending on location but amounts to about a month on average at these mid-latitudes.

In this observing project, you will explore how the orientation of the Earth's axis relative to its orbital plane produces the seasons. You will also investigate how the times and locations of sunrise and sunset change with the seasons, thereby affecting the length of daylight, and how these seasonal changes vary with latitude on the Earth.

A. Orientation of Earth's Rotation Axis Through the Year

In this section, you will have the opportunity to observe the Earth from several unique perspectives that demonstrate the cause of the seasons.

1. Launch *Starry Night Enthusiast*™.

2. Open the **Favourites** pane and select **Observing Projects > Seasons > Hovering Over South Pole**. If the Earth does not show initially, move the view slightly with the **Hand Tool** and click **File > Revert**.

This view is from a position hovering in space 45,000 kilometers above Earth's surface, almost directly over the south pole on March 21, 2006. The time is noon, 12:00:00 PM, along the prime meridian, the red line extending from the south pole on the Earth's image. On this date, the illumination from the Sun can be seen to reach just to the south pole, as expected for this date, the beginning of autumn in the southern hemisphere. The gaze direction is along the rotation axis of the Earth toward the north celestial pole, which is hidden behind the Earth from this location. The north ecliptic pole, the point perpendicular to the orbital plane of the Earth is labeled, and the ecliptic coordinate grid, based on the reference plane of the Earth's orbit, is displayed against the background sky. The displacement of the ecliptic pole away from the Earth's spin axis indicates the tilt of this axis, at an angle that can be estimated from the ecliptic coordinate grid.

> **Question 1.** What is the tilt of the Earth's axis to the perpendicular to the ecliptic plane, measured from the ecliptic coordinate grid?

> 3. [Optional] Use the **location scroller** to uncover the north celestial pole and verify its position. Then select **File > Revert**.
>
> 4. **Run time forward** for at least one year of simulated time and note the orientation of the Earth's axis with respect to the stars.

This viewing location is fixed with respect to the center of the Earth, so as the Earth orbits the Sun, the Sun (off the screen in this view), moving along the ecliptic in the sky, appears to move around the Earth. The stars remain stationary because the motion of the Earth in its orbit is insignificant compared to the distance to the background stars.

You can follow the progress of the Sun by watching the movement of the illuminated portion of the Earth's surface. Because the initial time is noon on the Earth's prime meridian and the time interval in the animation is one solar day, all subsequent views are at noon on the prime meridian. The prime meridian therefore maintains its direction toward the Sun, giving the appearance that the Earth is rotating. In fact, this appearance is the result of the difference between a solar day and a sidereal day, as you can investigate by changing the time interval to 1 sidereal day.

> 5. Change the **Time Flow Rate** to **1 sidereal day** and **Run time forward**.

Now, the Earth remains stationary in the view with respect to the background stars, but the sunlit hemisphere moves around the image as the Sun's position with respect to the background changes as the Earth orbits it. You can also see the illuminated region moving north and south as the Sun moves above and below the equatorial plane to bring more or less direct sunlight to mid-latitudes. In particular, you can see how the south polar region is alternately the "land of the midnight sun" in the southern summertime and the "land of 24-hour darkness" in the southern wintertime.

> **Question 2.** Toward which point in the sky is the north pole of the Earth's rotation axis directed throughout the year?

> 6. Select **File > Revert**.

> **Question 3.** In which direction on the screen (i.e., to the right, left, bottom, or top) is a) the Sun, and b) the north ecliptic pole?
>
> **Question 4.** Toward which direction on the screen is the north celestial pole tilted with respect to the north ecliptic pole?

7. **Increase current elevation** to about **0.001 AU** above the Earth and then use the **Angular Measurement Tool** to measure the angular separation between the Earth and the north ecliptic pole. In this view, the center of the Earth is aligned with the rotation axis of the Earth. Thus, this measurement indicates the amount by which the Earth's rotational axis is tilted from the perpendicular with respect to its orbital plane, the point marked by the north ecliptic pole.

Question 5. How far is the Earth's rotation axis tilted from the perpendicular with respect to its orbital plane?

We can now look at the Earth-Sun geometry from a different viewpoint, looking along the perpendicular to the ecliptic plane, that is, the Earth's orbital plane.

8. Open the **Favourites** pane and select **Observing Projects > Seasons > Along Ecliptic Axis**. If the Earth does not appear immediately, move the view slightly with the **Hand Tool** and click **File > Revert**.

This view is nearly identical to the previous one, except that the observing location has been shifted so that the line of sight is through the center of the Earth toward the north ecliptic pole, which is hidden behind the Earth's image. The south pole now points at an angle to the perpendicular to the Earth's orbital plane, the ecliptic plane.

9. [Optional] If you choose, use the **location scroller** to shift the point of view in order to identify the north ecliptic pole. Then select **File > Revert**.

Question 6. Toward which direction on the screen is a) the Sun and b) the north celestial pole?

10. **Zoom in** until the Earth almost fills the view.

Note the position of the terminator, the boundary between day and night on the Earth.

Question 7. How would you describe the position of the terminator with respect to the south pole of the Earth on this date?

We can now open a second window to display the position of the Earth at various positions in its orbit with respect to the Sun while observing the illumination from the Sun on the Earth at these positions in the original window.

11. Select **File > Revert**.
12. Select **File > New** to open a second *Starry Night Enthusiast*™ window.
13. In this new window, open the **Favourites** pane and select **Observing Projects > Seasons > Earth's Orbit**. If the **north ecliptic pole** does not appear at the center of the view, move the view slightly with the **hand tool** and select **File > Revert** from the menu.

This view shows the position of the Earth in its orbit around the Sun on March 21, 2006, from a position above the Sun. The line of sight is through the center of the Sun toward the north ecliptic pole, in other words, along the axis of the ecliptic, perpendicular to the Earth's orbital plane.

14. Position the **hand tool** over the object at the north ecliptic pole and verify from the **HUD** that the object at that location is the Sun.

Question 8. From the position of the Earth in its orbit, in which direction on the screen is a) the Sun and b) the north celestial pole?

15. Resize this window to about half its original width and move it to the left side of the screen.
16. If the toolbar has disappeared from the view, select **View > Show Toolbar** from the menu.
17. Select **Window > Along Ecliptic Axis**. Then, resize this window to about half its original width and move it to the right side of the screen so that both windows are visible, side-by-side. If the toolbar has disappeared from the view, select **View > Show Toolbar** from the menu.

With two windows open simultaneously, you can watch both the position of the Earth in its orbit in the first window and the resulting illumination on the southern hemisphere of the Earth in the second window.

18. Select **Window > Earth's Orbit** to reactivate this view and press the **M** key three times to advance the **Date** in the toolbar by 3 months, to **June 21, 2006,** and watch the Earth move in its orbit.
19. Select **Window > Along Ecliptic Axis** to reactivate this view and press the **M** key three times to change the **Date** in the toolbar to the same date as for the orbit window, **June 21, 2006.** This now shows the illumination of the Earth's hemisphere when the Earth is in the position shown in the other view.

Question 9. Which pole of the Earth is tilted toward the Sun on this date?

20. **Zoom in** so that the Earth's image almost fills the view.
21. Change the **Time Flow Rate** to **10 minutes** and **Run time forward.**
22. After several days of simulated time have passed, Select **Edit > Undo Time Flow.**

Question 10. Is the south pole in light or darkness on this date?

Question 11. What season is it in the southern hemisphere?

Question 12. Study the position of the terminator on the image of the Earth.

    a) At which parallel of latitude does the terminator cross the prime meridian on this date?

    b) If the Earth's rotation axis were perpendicular (90°) to its orbital plane, the terminator would remain in position across the south pole. How far from perpendicular is the rotation axis of Earth tilted with respect to its orbital plane based on your result in part a) of this question?

c) How does your answer to part b) compare to the measurement of the angular separation between the Earth's axis and the north ecliptic pole that you made earlier? Account for any differences.

23. **Zoom out** to the maximum field of view.

24. Select **Window > Earth's Orbit** to reactivate this view and press the **M** key three times to change the **Date** in the toolbar to **September 21, 2006**, thereby moving the Earth along in its orbit by another quarter revolution.

25. Select **Window > Along Ecliptic Axis** to reactivate this view and press the **M** key three times to change the **Date** in the toolbar to **September 21, 2006**.

Question 13. How would you describe the position of the terminator with respect to the south pole of the Earth on this date?

26. Select **Window > Earth's Orbit** to reactivate this view and press the **M** key three times to change the **Date** in the toolbar to **December 21, 2006**.

27. Select **Window > Along Ecliptic Axis** to reactivate this view and press the **M** key three times to change the **Date** in the toolbar to **December 21, 2006**.

Question 14. Which pole of the Earth is tilted toward the Sun on this date?

28. **Zoom in** so that the Earth's image almost fills the view.

29. Change the **Time Flow Rate** to **10 minutes**.

30. **Run time forward** for several days and then select **Edit > Undo Time Flow**.

Question 15. Is the south pole in light or darkness on this date?

Question 16. What season is it in the southern hemisphere?

Question 17. Study the position of the terminator on the image of the Earth. At which parallel of latitude does the terminator cross the prime meridian on this date?

31. **Zoom out** to the maximum field of view.

32. Select **File > Close** in either of the *Starry Night Enthusiast*™ windows and click **Don't Save** in the message box that appears. This will reactivate the remaining *Starry Night Enthusiast*™ window. Maximize this window.

In this section, you have examined the illumination of the Earth by the Sun at four positions of the Earth in its orbit, thereby demonstrating the reason for seasonal variations because of the varying angle of illumination of sunlight on the Earth's surface and the resulting variable heating.

B. Equinoxes and Solstices

In this section, you will watch the illumination by the Sun of the Earth as it moves in its orbit. The view is locked onto the Earth and thus the Sun will appear to move along the ecliptic as time advances.

> 33. Select **Favourites > Observing Projects > Seasons > Equinoxes and Solstices.**

This view on March 21, 2006, is from a position 80,000 kilometers above the surface of the Earth over the equator, looking along the plane of Earth's orbit, the ecliptic plane. The time is midnight, 12:00:00 AM. The ecliptic, a bright green line marked in intervals of months, shows the edge-on view of this plane. This is the path taken by the Sun in our sky, as seen from Earth. The tilt of the Earth's axis from the perpendicular to the ecliptic plane is evident from the orientation of the north and south poles. This tilt can also be seen between the ecliptic and the Earth's equator, which is projected onto the celestial sphere as the celestial equator (the red line on the background sky). On this date, March 21, 2006, the Sun, which, from the present viewing location is hidden behind the Earth, is at one of the positions where the celestial equator intersects the ecliptic, known as the **vernal** or **spring equinox.**

> 34. You can verify this by holding down the **Shift** key to change the **Hand Tool** to the **Location Scroller** and use it to move the Earth out of the way, and then use **File > Revert** to restore the original view.

What is immediately apparent at this time is that the visible hemisphere of Earth from our viewpoint is in total darkness, with no portion illuminated by the Sun. Observers on both poles would see the Sun on their horizon. Any observer at other than the poles will be in sunlight for one-half of the rotation period, and daylight will last for 12 hours. This geometry leads to the name **equinox**, which indicates equal day and night on this date and on the equivalent date 6 months later when the Sun crosses the celestial equator again from our earthbound viewpoint, at **autumnal equinox**, the beginning of autumn in the northern hemisphere.

> The vernal equinox and autumnal equinox are illustrated in Fig. 2–15 of Freedman, Geller, and Kaufmann, *Universe,* 9th Ed.

> **Question 18.** If you were on the surface of Earth on the equator at noon on this date, where in the sky would you expect to see the Sun?

One important plane associated with the Earth is the plane that contains both the spin axis of the Earth and the axis perpendicular to the ecliptic plane that joins both north and south ecliptic poles. In the view currently displayed, the ecliptic poles are out of the view toward the top and the bottom of the view since the view is along the plane of the ecliptic. Consequently, the plane that contains both the spin axis of the Earth and the ecliptic axis in this view passes through the center of the Earth and is oriented parallel to the plane of the current view. This is shown in Figure 1 below. In this figure, the plane of the paper is the ecliptic plane. The view is from the direction of the north ecliptic pole, perpendicular to this plane. The Earth is shown at two positions in its orbit separated by about three months. The direction to the north ecliptic pole on each sketch of the Earth is perpendicular to the plane of the paper and indicated on the depiction of the Earth as a black dot. The north pole of the spin axis of the Earth is indicated by a black diamond. It is offset from the north ecliptic pole because of the tilt of the Earth's axis to the ecliptic (this offset is exaggerated in the figure). The reference plane containing both the spin axis of the Earth and the ecliptic poles is perpendicular to the paper in the figure, and its orientation at each position of the Earth is indicated by the dotted line that contains both the north pole and the north ecliptic pole.

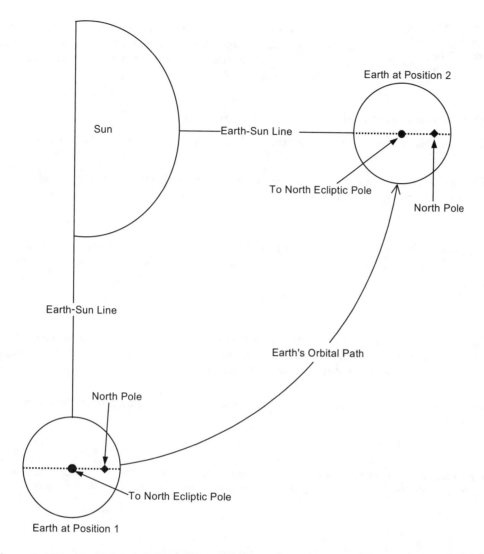

Figure 1. Plane Containing Ecliptic Poles and Spin Axis of Earth at Two Positions in Earth's Orbit

Question 19. On the date shown in the view window, March 21, is the Earth-Sun line perpendicular or parallel to the reference plane containing the Earth's spin axis and the ecliptic polar axis?

You can now advance the date and watch the Sun move around the ecliptic from west to east, thereby changing the illumination on the Earth. The Time Flow Rate is set to 1 sidereal day. Thus, the Earth will appear to remain fixed with respect to the background stars as time advances. The local time at the point directly below the observer will change because of the difference between sidereal time and solar time.

35. **Run time forward** in 1-sidereal-day time steps to **June 21, 2006,** to watch the Sun move from behind the Earth and begin to illuminate a different region of the Earth.

36. **Zoom in** on the Earth to a field of view of about **25°** to examine the illumination of the Earth on this date.

The solstices are illustrated in Fig. 2–15 of Freedman, Geller, and Kaufmann, *Universe*, 9th Ed.

Note that the most direct sunlight now falls on a region north of the equator. The Sun is at a position called **Summer Solstice**, so named because it has reached its highest angle in the sky from northern latitudes, the word "solstice" meaning that the Sun stands still in this north-south annual motion.

Question 20. Which pole of the Earth is tilted toward the Sun on this date?

Question 21. From this illumination pattern on the Earth, what season is it in the southern hemisphere?

Question 22. On this date, is the Earth-Sun line perpendicular or parallel to the plane containing both the spin axis of Earth and the ecliptic polar axis?

37. Keep the expanded view of Earth and **Run time forward** to **September 21, 2006,** watching the change in solar illumination as the Sun progresses around behind your observing position to the autumnal equinox.

Note that, on this date, the hemisphere of the Earth facing the observer is fully illuminated, including both poles. Again, you can see that the length of the day will be the same for all observers on Earth (except those at either poles) and that sunlight is falling directly down on the equator.

Question 23. Where will the Sun appear in the sky for observers on the south pole of the Earth on this date?

38. **Run time forward** in sidereal time to **December 21, 2006,** the date of the winter solstice. The Sun has now moved to the far right of its orbit when viewed from the present observing position.

Question 24. Which pole of the Earth is tilted toward the Sun on this date?

Question 25. What season is it in the southern hemisphere on this date?

Question 26. In which hemisphere would you be if the Sun were to pass through your zenith at midday on this date?

39. [Optional] It is interesting to **Run time forward** for another year or two while watching the pattern of illumination change over the whole Earth.

You have watched the north-south movement of the area of illumination across the Earth as the Earth moves around the Sun in its orbit. It is this that produces the varying heating and cooling on the surface, leading to summer and winter seasons at mid-latitudes on the Earth.

C. The Seasonal Change in the Position of the Sun

The Sun is at the center of the ecliptic plane and appears to move north and south in the sky, varying with the seasons as seen from Earth because the Earth's axis is tilted with respect to this plane. This section will demonstrate this motion from a viewpoint on the equator.

40. Select **Favourites > Observing Projects > Seasons > View from Earth.**

The gaze is toward the zenith near noon on March 21, 2007, from a viewing location on Earth at the longitude of the Greenwich meridian and the latitude of the equator. On this date, at this time, the Sun can be seen at the point of intersection of the ecliptic and the celestial equator and is labeled for convenience.

41. Select **View > Local Meridian.**

> **Question 27.** According to the scale on the local meridian, what is the approximate altitude of the Sun as seen from the equator at noon on this date?

42. Use the **M** key to advance the **Date** to **June 21, 2006.**

> **Question 28.** What is the altitude of the Sun near noon from the equator on this date?
>
> **Question 29.** How far, in degrees, is the Sun from the zenith (90° altitude)?

43. Use the **M** key to advance the **Date** to **September 21, 2006.** Note the altitude of the Sun from the local meridian.
44. Use the **M** key to advance the **Date** to **December 31, 2006,** and note the altitude of the Sun from local meridian.

> **Question 30.** As seen from the equator of Earth, near noon, what is the altitude of the Sun on a) September 21 and b) December 21?

D. The Position and Time of Sunrise Through the Year

In this part of the project, you will see how the location of sunrise on the horizon changes throughout the year at one location in the northern hemisphere and how this variation is matched by changes in the time of sunrise.

45. Select **Favourites > Observing Projects > Seasons > Sunrise.**

Starry Night Enthusiast™ shows the view looking east (Gaze azimuth of 90°) at sunrise from New York City on March 21, 2007. The horizon is replaced in this view by a grey line. In the following sequence, you will observe the position and time of sunrise through the year.

46. Select **File > Preferences** (Select **Starry Night Enthusiast > Preferences** on a Macintosh system). Choose **Cursor Tracking (HUD)** in the Preferences dialog and select only **Azimuth** and **Name** from the **Show** list.

47. Click the **seconds** field of the **Time** in the toolbar and use the "+" and "−" keys (or the cursor up and down keys) on the keyboard to adjust the time so that the upper edge of the Sun just touches the horizon. Enter the time of sunrise in Data Table 1.

48. Position the **Hand Tool** over the Sun and note the **Azimuth** of the Sun from the **HUD**. Round this value off to the nearest ¹/₂ deg. Record this value under the **Azimuth** column in Data Table 1. Calculate the position angle between due East and the Sun at sunrise by subtracting this value from 90° (due east) and enter the result, with the proper sign, into the Data Table. The result of this calculation will be positive when the Sun rises to the north of east and negative when it rises to the south of east.

Data Table 1. Time and Azimuth of Sunrise from New York Throughout the Year

Date	Time of Sunrise	Azimuth (°)	90° − Azimuth (°)
March 21, 2007			
April 21, 2007			
May 21, 2007			
June 21, 2007			
July 21, 2007			
August 21, 2007			
September 21, 2007			
October 21, 2007			
November 21, 2007			
December 21, 2007			
January 21, 2008			
February 21, 2008			
March 21, 2008			

49. Use the **M** keyboard shortcut to increment the **Date** by one calendar month.

50. Use the "+" and "−" keys on the keyboard to adjust the **minutes** and **seconds** of the **Time** in the toolbar until the upper edge of the Sun just touches the horizon line. Record the time of sunrise in Data Table 1.

51. Position the **Hand Tool** over the Sun and note its **Azimuth** from the **HUD** into Data Table 1.

52. Calculate the difference between the Sun's azimuth and due east and record the result of the calculation in Data Table 1.

53. To enter the data for the remaining dates, repeat the last four instruction steps for each date in the Data Table.

Now you can use your recorded measurements to plot the sunrise position (represented by "90° – Azimuth" in the data table above) graphically against the date on the graph template in Figure 2 below.

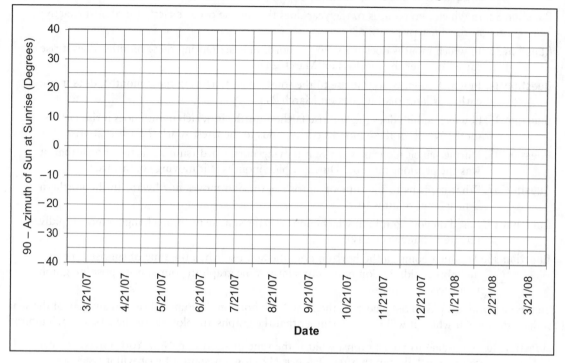

Figure 2. Graph Template for Plotting Sunrise Position from Due East Through the Year

Plot a second graph, this time showing the sunrise time against the date in the year. You can use the graph template in Figure 3 below to plot this data.

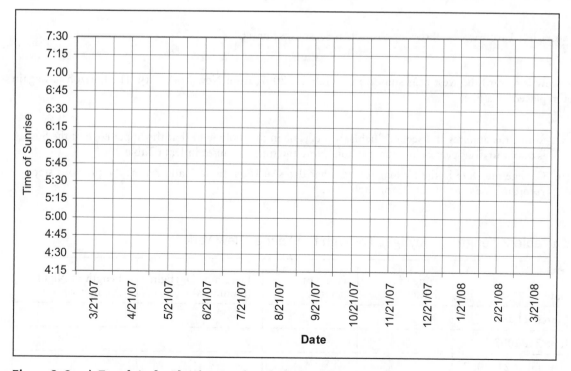

Figure 3. Graph Template for Plotting Sunrise Time Through the Year

Question 31. What is the shape of the graph of sunrise position as a function of time of the year (e.g., straight line, sinusoidal [i.e., S-wave] curve, etc.)?

Question 32. In which two months of the year does the sunrise occur exactly (or almost exactly) at the due east position?

Question 33. In which months does it rise to the north of east? During these months, does it rise earlier or later than it does on March 21?

Question 34. In which months does it rise to the south of east? During these months, does it rise earlier or later than it does on March 21?

Question 35. In which month does the Sun rise farthest north? In which month does it rise farthest south? Which of these months corresponds to the latest sunrise? Earliest sunrise?

Question 36. During which months is the Sun moving northward (sunrise farther north than it was the previous month)? During which months is it moving southward?

Question 37. What happens to the time of sunrise as the Sun moves northward in the northern hemisphere? Southward?

Question 38. Based on your graph, in which months is the sunrise position changing most rapidly (greatest change per day)?

Question 39. At which point on the horizon does the Sun rise when the time of sunrise is changing most rapidly? (Compare your sunrise time graph to your sunrise position graph to answer this question.)

You can now use your graphs to tell where on the New York horizon the Sun will rise on any day of the year and estimate the time at which it will rise. Remember that the graphs are plotted for the 21st of each month.

Question 40. According to your graphs, what is the time of sunrise in New York for November 5; i.e., one half month after October 21? How far south of east will it rise?

E. Change in the Length of the Daylight Through the Year

Having observed sunrise throughout the year from the latitude of New York, you can now make observations of sunset from the same location in this section of the project.

54. Select **Favourites > Observing Projects > Seasons > Sunset**.

The view is toward the west near sunset on March 21, 2007, from New York City. The horizon has again been replaced by a line.

55. Adjust the **minutes** and **seconds** fields of the **Time** in the toolbar so that the upper edge of the Sun is about to disappear below the horizon. Note the time of sunset in Data Table 2.

56. Change the **Date** in the toolbar to the other dates in Data Table 2 and repeat the previous step, recording the time of sunset for each date.

Data Table 2. Length of Daylight Through the Year at New York

Date	Time of Sunset	Time of Sunrise	Length of Daylight (hh:mm:ss)	Length of Night (hh:mm:ss)
March 21, 2007				
June 21, 2007				
September 21, 2007				
December 21, 2007				

57. For the sunrise times in Data Table 2, use the sunrise times that you measured and recorded in Data Table 1 in the previous section.

58. Use the times of sunset and sunrise to calculate the duration of daylight for each of the dates in Data Table 2. You can also calculate the length of the night, which will be equal to 24 hours minus the duration of daylight.

Question 41. What approximate relationship do you notice between the lengths of the day and night for June 21 and December 21? (For instance, how does the length of the day on June 21 compare to the length of the night on December 21?)

Question 42. Based on your values in Data Tables 1 and 2, what happens to the length of the day as the rising or setting point of the Sun moves northward along the horizon? Southward?

Question 43. We have kept the time set to standard time throughout this exercise, but in fact most areas use Daylight Savings Time (or Daylight Time) between April and September. If Daylight Time were in effect, would it have any effect on the length of the day or night? (Remember that we set our clocks one hour ahead when daylight time begins, so BOTH sunrise AND sunset takes place one hour later than they would if we had left the clocks at standard time.) What then does the phrase "Daylight Savings" mean?

F. Duration of Daylight at the Equator

As you have seen, summer brings longer days and shorter nights to mid-northern latitudes. Conversely, in winter, the days are shorter and nights longer. The length of the day over the course of the seasons depends, however, on one's latitude on Earth.

59. Select **Favourites > Observing Projects > Seasons > Daylight at Equator.**

The view is centered on the Sun on March 21, 2007, and the field of view decreased to provide a close-up of the Sun at around the time of sunrise as seen from the equator on Earth.

60. Adjust the **Time** in the toolbar so that the upper limb of the Sun just touches horizon line. Note the time of sunrise for this latitude and record it in Data Table 3.

61. Note the **Azimuth** of the Sun at sunrise from the **HUD**. Subtract the azimuth from 90° and record the result of this calculation including the sign under the column **90° – Azimuth** in Data Table 3. This gives you the sunrise position relative to east with positive values indicating a position north of east and negative values a position south of east.

62. Click the AM field of the **Time** in the toolbar to change it to PM. Since the view is locked on the Sun, the view changes to the west near the time of sunset. Adjust the **Time** in the toolbar until the Sun is about to disappear below the horizon line. Record the time of sunset in Data Table 3.

63. Find the time of sunrise, azimuth of the Sun at sunrise, and the time of sunset for each of the other dates listed in Data Table 3 using the techniques of the previous three steps. Record these data in the Data Table.

64. Calculate the length of the day and of the night at the equator for each of the dates.

Data Table 3. Length of Daylight through the Year at the Equator of Earth

Date	Time of Sunrise	Time of Sunset	Length of Daylight (hh:mm:ss)	Sun's Azimuth at Sunrise (°)	90° − Sun's Azimuth at Sunrise (°)
March 21, 2007					
June 21, 2007					
September 21, 2007					
December 21, 2007					

Question 44. What can you say about the length of daylight through the year at the equator?

Question 45. How does the position of the Sun at sunrise relative to due east (azimuth 90°) compare to the tilt of the Earth's spin axis to the ecliptic axis?

In the next sequence, you will observe sunrise and sunset from the north pole.

65. Select **Favourites > Observing Projects > Seasons > Daylight at North Pole.**

The view is centered on the Sun. From the north pole on March 15, 2007, at 6:00:00 AM, the Sun is still below the eastern horizon.

66. With the **Time Flow Rate** set to **1 hour, Run time forward** and then **Stop Time** as soon as the Sun touches the horizon line. Note the date on which the Sun rises as seen from the north pole during 2007.

67. Change the **Time Flow Rate** to **4 hours** and, with the view locked on the Sun, **Run time forward.**

68. **Stop Time** when the Sun nears the horizon again and then **Step time forward** to determine the date of sunset in 2007 as seen from the north pole.

Question 46. How long does daylight last as seen from the north pole?

G. Conclusions

You have explored the cause of the seasons and observed times of sunrise and sunset and the relative lengths of day and night in winter and summer. The tilt of the Earth's axis is the main contributor to the seasonal effects on mid-latitudes, where the changing inclination of sunlight to the Earth's surface changes the efficiency of heating. The extended daylight in summertime also provides longer growing times for crops in summer.

These seasonal effects are also present on other planets. Uranus, with its spin axis almost in the plane of its orbit, will show the most pronounced seasonal variations in heating and cooling.

The Analemma

8

The Sun reaches its highest point in the sky every day when it crosses an observer's celestial meridian. This occurs close to midday every day. The **celestial meridian,** often called simply the meridian, is an imaginary line passing through the observer's zenith and the north and south celestial poles and is shared by all observers on the same line of longitude on Earth. From the northern hemisphere, this meridian passes vertically through the south point of the observer's horizon. In fact, all objects in the background sky reach their highest point on the observer's sky as they pass through the meridian. If we observe the Sun's position at twelve o'clock each day throughout the year however, we will rarely find it on the meridian. There are several reasons for the Sun being offset from the meridian at noon.

> The meridian is defined in Section 2–7 and illustrated in Figure 2–21 of Freedman, Geller, and Kaufmann, *Universe*, 9th Ed.

> Freedman, Geller, and Kaufmann, *Universe*, 9th Ed., Figure 2–23 shows a map of Earth's time zones.

The first reason is that the observer's location may not be at the center of the time zone. This produces a constant offset of the Sun from the observer's celestial meridian at noon. If time zone boundaries were ideal, this offset might be as large as half an hour. For political reasons, time zone boundaries sometimes differ from the ideal positions, resulting in offsets of close to an hour in some locations.

The second reason is that the Earth's orbit is elliptical. This means that both the Earth's orbital speed and its distance from the Sun vary through the year, as described by Kepler's second law. The varying speed and solar distance cause the apparent speed of the Sun past the background stars to change day by day.

> Section 4–4 of Freedman, Geller, and Kaufmann, *Universe*, 9th Ed., discusses Kepler's second law.

The third reason is that the Earth's spin axis is tilted with respect to the perpendicular to the plane of the Earth's orbit, the ecliptic plane. The tilt also causes a variation in the effective speed of motion of the Sun in a direction parallel to the Earth's equator. It is this tilt of the Earth's spin axis that also causes the variations of the seasons.

The combination of the Earth's elliptical orbit and its axial tilt causes a variable offset of the Sun from the meridian at midday, with the Sun being displaced east and west of the meridian in a cyclic pattern over the course of a year. At this same time, the tilt of the Earth's axis also causes the Sun to drift north and south of the equator as the seasons progress. The result is a figure-8 pattern for the Sun's midday position in our sky over the course of a year. This figure-8 pattern is called the **analemma,** and its shape is often depicted upon globes representing the Earth.

> Section 2–7 in Freedman, Geller, and Kaufmann, *Universe*, 9th Ed., discusses the effects of these deviations upon timekeeping.

The shape of the analemma was important when sundials were used to tell the time. The analemma provided corrections to sundial times, amounting to as much as 16 minutes at certain times of the year. The graph of the difference between the time of the Sun's passage through the meridian and midday, the time error of a simple sundial, is sometimes known as the **Equation of Time.** This observing project demonstrates the analemma and examines how Earth's axial tilt and elliptical orbit produce its shape.

A. The Shape and Size of the Analemma

The first simulation will view the analemma from Calgary, Alberta, Canada, a city that lies at a longitude of about 114° W, within the Mountain Standard Time zone. The standard meridian for this zone is 105° west of the Greenwich meridian. The Greenwich meridian, also known as the prime meridian, passes through London, England, and defines the zero point of longitude. Universal Time, a common standard time used throughout the world, is maintained at the Greenwich meridian, 0 hours UT being midnight on this meridian.

The Earth rotates through 360° in 24 hours, or 15° per hour. The standard time in the Mountain Standard Time zone (MST) is therefore 105/15 = 7 hours behind Universal Time.

In fact, Calgary is 9° west of the standard meridian for the Mountain Standard Time zone, which means that when a person on the standard meridian sees the Sun on the celestial meridian, a person in Calgary sees the Sun 9° east of the celestial meridian. We must therefore adjust the time to allow for this 9° offset in order to place the Sun on the meridian in Calgary on average throughout the year. At 15° per hour, this offset amounts to 36 minutes of time. Because Calgary is west of the standard meridian, the Sun crosses the meridian later, at 12:36 PM, MST.

Question 1. Boston, Massachusetts, is at longitude 71° 5' W in the Eastern Standard Time zone. The standard meridian for this zone is 75° W. On the basis of Boston's position in the time zone, where would you expect to find the Sun relative to the meridian at noon Eastern Standard Time, if a person at 75° W sees the Sun on the meridian?

Question 2. If a person at 75° W sees the Sun on the meridian at noon, at what time would a person in Boston see the Sun on the meridian?

> 1. Launch *Starry Night Enthusiast*™ and select **Favourites > Observing Projects > Analemma > Solar Time**.

The view shows the southern horizon from Calgary on December 21, 2006 AD, the date of winter solstice. The local meridian is displayed as a line from the south point on the horizon to the zenith, beyond the top of the screen. Time is set to 12:34:10 MST to place the Sun on the meridian for Calgary's specific position within the Mountain Standard Time zone on this date.

At this time of year in the northern hemisphere, the Sun is at a low angle above the southern horizon. A vertical pole acting as a sundial at this location would cast a long shadow away from the south point on the horizon, toward the north. You can now **Run time forward** in 3-day intervals to display the shape of the analemma on the sky.

> 2. With **Time Flow Rate** set at **3 days**, **Run time forward** and observe the change in position of the midday Sun at 3-day intervals through the year. Allow time to run for a year or two.
>
> 3. **Stop Time.**
>
> 4. **Tip:** Click an elevation button in the control panel to erase the analemma.
>
> 5. Select **View > Hide Daylight** and start time again to show the motion of the Sun against the background stars. Since the view direction is locked on to the meridian, the background stars appear to move as a consequence of the difference between the lengths of solar and sidereal days.

You can see that the position of the midday sun traces out a figure-8 pattern in a full year, climbing high in the sky in summer and returning to low elevation angles in winter. For a sundial to agree with our clocks, the Sun would have to be on the meridian at midday every day. It is apparent that this is rarely the case. The Sun is often quite far from this ideal position, leading to errors in timing from the use of a simple sundial.

B. The Tilt of the Earth's Spin Axis

The reason for the north-south motion of the Sun is the tilt of the Earth's axis to the ecliptic plane. It is this tilt that produces seasonal changes on the Earth. The maximum north-south excursion of the Sun, in degrees of declination, will be twice the tilt-angle of the Earth's spin axis to its orbital plane. We can measure this north-south excursion using the analemma figure.

6. Change the **Time Flow Rate** to **1 day** and use the time controls to move the Sun's position to the bottom of the analemma.

7. Open the **View** pane and clear the checkbox next to **Stars** under the **Stars** layer. This will hide the stars in the view. Then use the **Angular Measurement Tool** to measure the angle between the Sun and the top of the analemma. Note this angle in Data Table 1 below.

Data Table 1. Measuring the Analemma to Determine the Tilt of Earth's Rotation Axis

Angular Length of Analemma (° ' ")	
Tilt of Earth's Axis (Analemma length × 0.5)	

Question 3. What is the tilt angle of the Earth's spin axis relative to the perpendicular to its orbital plane?

C. Errors in Sundial Times

The analemma represents the Sun's midday position with respect to an observer's meridian throughout the year. If you needed to use a simple sundial such as the shadow of a vertical pole to tell the time (on a desert island as part of the "reality" television program *Survivor* maybe!), you would be in error on any given day by the time taken for the Sun to move through the angle represented by the difference between the analemma and the meridian. As you can see, your "clock" would be slow or fast, depending on whether the point on the analemma was east or west of the meridian, respectively. We can measure these errors, using the traced analemma pattern.

8. **Run time forward** until the Sun is at the widest point of the analemma on the east (left) side of the meridian. Record the **Date** of this greatest deviation of the Sun from the meridian in Data Table 2 below.

9. Measure the angular distance, θ, between the Sun and the meridian and convert this angle into degrees and fractions of a degree. To do so, divide the number of minutes of angle in the measurement by 60 and add the result to the number of degrees in the measurement. Note this angle in Data Table 2 below.

10. Repeat this measurement by moving the Sun to the equivalent position on the west side of the analemma and measure and record the error angle and the **Date** in Data Table 2.

The Earth rotates through 360° in almost 24 hours, the meridian moving across the background sky at a rate of 15° an hour. Thus, it takes 4 minutes of time for the meridian of the Earth to cover 1° of angle. You need to multiply the measured angle, θ, in degrees by 4 to calculate the sundial error in minutes of time.

Data Table 2. Maximum Sundial Errors

Date	Angular Distance of Sun from Meridian, θ (°)	Sundial Error, $\theta \times 4$ (minutes of time)

Question 4. What are the worst errors, in minutes of time, of the sundial for the two dates of maximum error?

Question 5. At what times of the year will the sundial be most inaccurate? (In practice, determination of the date of maximum error is not very precise.)

You will notice from the analemma that a sundial is accurate compared to our clocks on some days of the year, with the Sun on the meridian at noon.

11. Use the time flow controls in the toolbar to find the dates of the year when the sundial error is zero and note these dates.

Question 6. On what dates in the year will a sundial be most accurate?

D. The Analemma from Other Latitudes

The shape of the analemma will be the same from any position on Earth. You can test this hypothesis by moving to another location.

12. **Stop time and change the Date to June 21.** The specific year is not important.

13. Select **Options > Viewing Location...** and change the observing location to **São Paulo, Brazil.**

14. Turn off **Daylight Saving Time** if it is on and change the **Time** in the toolbar to **12:00:00 PM.**

15. Set the **Gaze** direction to due North.

16. Use the **Hand Tool** to drag the sky downward so that the Sun is at the bottom of the view and the meridian is vertical.

17. **Run time forward** to trace out the analemma from Brazil and **Stop Time** when the Sun has traced out the full pattern.

You will note that the shape of the analemma is the same as it was from Calgary. Measurements equivalent to those carried out above will confirm that it has exactly the same dimensions when observed from São Paulo. There is, however, one difference.

Question 7. What is different about the appearance and/or location of the analemma in the sky when viewed from São Paulo, Brazil, compared to that seen in Calgary? Why is this?

E. Analemma Details

In this section, we will explore in more detail how the tilt of the Earth's axis and its motion in an elliptical orbit around the Sun produce the sundial errors measured in the previous section. As the previous simulations have shown, the Sun sometimes runs slow and at other times fast compared to the time on clocks. This is because clocks are built to run at a steady rate, calibrated to what is called the **mean solar day**. The duration of a mean solar day is the time required for a hypothetical Sun, moving eastward throughout the year at a steady rate, to return to the same position in the sky. Sundials, on the other hand, measure the **apparent solar day**. The length of the apparent solar day is the time required for the *actual* Sun to return to the same position in the sky as seen from a particular location on Earth. Sundial errors, as demonstrated by the analemma, represent the degree to which the actual Sun lags behind or runs ahead of the hypothetical mean sun through the year. These errors, as mentioned above, are known as the Equation of Time.

The two effects that combine to produce this offset of the Sun from its mean position, day by day, are the ellipticity of the Earth's orbit and the tilt of the Earth's spin axis to its orbital plane. We can demonstrate the effect upon the analemma of each of these orbital parameters in turn by introducing a fictitious planet, which we will call Vulcan, into *Starry Night Enthusiast*™. Initially, we will create an orbit for Vulcan with the same properties as those of Earth's orbit. We can then edit these orbital properties in *Starry Night Enthusiast*™ to remove these effects individually and observe the resulting shape and size of the analemma.

18. Select **File > Revert**.

19. Select **File > New Asteroid Orbiting Sun . . .** from the menu. This will open a dialog window in which you will create the fictitious planet Vulcan.

20. At the top of the dialog window select **Planet** from the drop box. Then type **"Vulcan"** in the edit box.

21. Select the **Orbital Elements** tab in the lower section of the window and type the values of the parameters on this page into the appropriate boxes, as follows:

Style:	Near-circular
Ref Plane:	Ecliptic 2000
Mean distance (a):	1.0 AU
Eccentricity (e):	0.0167058

22. Click the **Other Settings** tab in the dialog window and set the following values:

Rotation rate:	1.002738	rotations per day
Pole declination:	90.0	degrees

23. Minimize, but do not close, the **Planet: Vulcan** dialog window.

24. Open the **Find** pane and click on the icon of the magnifying glass in the search box at the top of the pane. Select **Orbiting Objects** from the list that appears. The list in the **Find** pane should now display an entry for Vulcan.

25. Click the menu button for **Vulcan** and select **Go There**.

26. Click and hold the **Decrease Current Elevation** button in the toolbar to drop to the surface of this imaginary planet.

27. Set the **Gaze** direction to the South.

28. Check that the **Date** is **December 21, 2006** AD, and adjust the **Time** in the toolbar so that the Sun is on the meridian.

29. Use the **Hand Tool** to drag the sky downward so that the Sun is near to the bottom of the view.

You are in the northern hemisphere of the fictitious planet Vulcan, looking south at the Sun on the meridian. Vulcan's orbit around the Sun is virtually identical to that of the Earth.

> 30. **Run time forward** and observe the analemma.

You will see that this analemma is the same general shape as that seen from Earth, having asymmetrical north and south lobes and an axis that is tilted slightly with respect to the meridian.

You can now adjust the ellipticity of the orbit of this planet to make its orbit circular.

> 31. With time continuing to flow, restore the **Planet: Vulcan** dialog window.
> 32. Click the **Orbital Elements** tab and set the **Eccentricity (e)** to 0. Minimize the **Planet: Vulcan** window.

This step has reduced Vulcan's orbit to a circle. The resulting analemma is the result ONLY of the effect of the tilt of the planet's axis to its orbital plane, in this case, the Ecliptic Plane.

> **Question 8.** What effect does eliminating the eccentricity from Vulcan's orbit have on the appearance of the analemma?

You can now adjust the orbital parameters of Vulcan to demonstrate the effect of ellipticity only, without the tilt of its spin axis to the perpendicular to its orbital plane. The easiest way to accomplish this is to move the orbital plane of the planet to correspond to its equator. This will tilt Vulcan's orbital plane by 23.4° and move it away from the ecliptic plane. You will also need to restore the eccentricity of Vulcan's orbit.

> 33. Restore the **Planet: Vulcan** window and under the **Orbital Elements** tab, change the **Eccentricity (e)** back to 0.0167058. Then change the **Ref Plane** to **Earth Equatorial 2000**.
> 34. Minimize the **Planet: Vulcan** window and observe the analemma

Vulcan now has an elliptical orbit again but its spin axis is perpendicular to its orbital plane.

> **Question 9.** Describe the analemma for Vulcan when its orbit is elliptical but its spin axis is perpendicular to its orbital plane. How do you account for the shape of the analemma under these orbital conditions?

> 35. Restore the **Planet: Vulcan** window and change the **Eccentricity (e)** to 0 once again. Then minimize the **Planet: Vulcan** window and observe the Sun in the view.

Vulcan is now moving in a circular orbit and rotating once per day on an axis perpendicular to the plane of its orbit.

> **Question 10.** Describe the "analemma" as it appears on Vulcan under these orbital conditions. Explain your observations.

> **Question 11.** How accurate would a sundial be on the planet Vulcan as it is currently configured?

> **Question 12.** How would the two types of solar day, mean and apparent, compare on Vulcan as it is configured in this view?

36. Close the **Planet: Vulcan** window and click the **Don't Save** button in the message box that appears.

F. Sunrise Time at the Winter Solstice

The shape of Earth's analemma can help to answer the following question often asked about sunrise time: "If the Sun is at its farthest south on the shortest day of the year, at the time of solstice on December 21, why is this *not* the date of latest sunrise?" In fact, the date of latest sunrise is nearer to the end of the year! We will define sunrise as the moment when the upper limb or edge of the Sun just touches the horizon. Since this time will depend to some extent on the local horizon, the official definition assumes that the observer is at sea level. In order to simulate this situation, some of the following sequences use a line to represent a flat horizon.

37. Select **Favourites > Observing Projects > Analemma > Morning Analemma**.

The view is from Calgary looking southeast just after sunrise on December 21, 2006.

38. **Run time forward.**

In contrast to the view of the analemma at midday, this pattern is now tilted at a distinct angle to the horizon and is not perpendicular to it. Furthermore, the lowest point of the analemma pattern is no longer the position of the Sun at the winter solstice. The Sun will rise at the latest time of the day on the date when it is at its lowest point on this tilted analemma. You can demonstrate this with the following simulation.

39. Select **Favourites > Observing Projects > Analemma > Latest Sunrise**.

The view is from Calgary on December 21, 2006, just before sunrise. The gaze is centered on the Sun and the full horizon has been removed from the display and replaced by a line.

40. Adjust the minutes and seconds of the **Time** in the toolbar to the moment of sunrise, when the upper edge of the Sun just touches the horizon line. Record the time of sunrise in Data Table 3 below.
41. Change the **Date** in the toolbar to **December 23**. Once again, adjust the **Time** in the toolbar to the precise moment of sunrise, and record this time in Data Table 3. Repeat this step for the rest of the dates in Data Table 3.

Data Table 3. Time of Sunrise near the Winter Solstice

Date	Dec. 21	Dec. 23	Dec. 25	Dec. 27	Dec. 29	Dec. 31	Jan. 2	Jan. 4
Time of Sunrise								

Question 13. What is the date of the latest sunrise in the winter of 2006–2007 AD?

You can readily see that the date of latest sunrise is many days later than the date of the winter solstice. The motion of the Sun because of Earth's orbital parameters, as represented in the shape of the analemma, affects the sunrise time. You can demonstrate this by running the analemma on the date of the latest sunrise.

42. Select **Favourites > Observing Projects > Analemma > Morning Analemma**.

43. Set the **Date** and **Time** in the toolbar to the latest sunrise of the year.

44. Using the keyboard shortcuts **D** and **Shift-D** to step forward and backward in time by 1 day, respectively, to trace out the lower region of the analemma at this time.

Question 14. About how many days after the winter solstice does the latest sunrise occur?

Question 15. By how many minutes and seconds is this sunrise later than that at the solstice?

G. [OPTIONAL] Sunset Time at the Winter Solstice

It would be a useful exercise for you to use the techniques of the previous section to demonstrate this equivalent effect upon sunset times.

45. Use the views named **Evening Analemma** and **Earliest Sunset** in **Favourites > Observing Projects > Analemma** and follow the ideas presented in the sequences of the previous section to explore sunset near the winter solstice.

Question 16. What is the date of earliest sunset around the time of the winter solstice?

There will be similar time shifts of the earliest sunrise and latest sunset around the time of longest day in the summer.

Question 17. Do you think that these shifts of earliest sunrise and latest sunset around the date of the longest day of the year at summer solstice would be equal to, less than, or greater than the shifts of latest sunrise and earliest sunset around the time of winter solstice? Use the shape of the analemma to support your answer.

H. Conclusions

In this project, you have demonstrated the combined effects of the Earth's elliptical orbit and the tilt of the Earth's spin axis to the perpendicular to this orbit on the position of the Sun at midday through the course of the year and have seen this effect displayed as the analemma on the sky. You have measured the possible errors in a simple sundial because of these effects. These errors, when plotted for the whole year, are known as the Equation of Time. This "equation" is often displayed on modern sundials. You have also used the analemma to demonstrate that the date of latest sunrise in the winter is not the date of the shortest day of the year at the winter solstice.

Precession and Nutation of the Earth

9

As the Earth orbits the Sun in the ecliptic plane, it also rotates in space with a period of one sidereal day around an axis that is inclined to this orbital plane. This angle of inclination varies over very long periods but is about $23\frac{1}{2}°$ at the present time. If the Earth's shape were a perfect sphere, its rotation would be undisturbed and the direction of the axis would remain fixed in space. However, the Earth's rapid rotation, particularly during its formative stages, has distorted it into a flattened shape known as an oblate spheroid in which its equatorial diameter is now 43 km longer than its polar diameter. In addition to the direct gravitational forces that act between the centers of the Earth, the Moon, and the Sun to maintain their respective orbital motions, this slightly distorted Earth now becomes subject to the influence of differential gravitational forces from the Moon and the Sun.

These differential forces act on Earth's equatorial bulge to cause its spin axis to move in a conical motion called **precession**. Since the Moon and Sun remain close to the ecliptic plane, they generate an average force on the bulge at an angle of $23\frac{1}{2}°$ to the equatorial plane. This disturbs the simple spinning motion of the Earth.

> These differential forces also cause the tides in the Earth's oceans by acting with slightly different strengths on equivalent masses on either side of Earth. These tidal effects are discussed in Section 4–8 of Freedman, Geller, and Kaufmann, *Universe*, 9th Ed.

This slow coning precessional motion of the axis is equivalent to that of a child's spinning top. When a spinning top is placed on the ground with its axis inclined to the vertical, there are out-of-balance forces acting on the top that are attempting to cause it to fall over. Indeed, if it were not spinning, the top would fall over under gravity. In both cases, spinning top and rotating Earth, out-of-balance forces cause classical precession of the spin axis. The actual dynamics are relatively simple but require knowledge of the physics of rotational motion that is beyond the scope of this book.

The consequences of this precessional motion upon life on Earth are very small and affect almost no one. However, they do affect the orientations of the equatorial plane and the spin axis of Earth. The extensions of the Earth's equator and of the spin axis form the basis for a system of coordinates, **right ascension** and **declination,** used by astronomers to define positions of objects in space. Precession therefore causes the right ascension and declination of each star to change slowly and continuously over time as the direction of the Earth's rotation axis changes. Star atlases have to be specifically labeled with the epoch for

> Section 2–6 in Freedman, Geller, and Kaufmann, *Universe*, 9th Ed. discusses epochs. Box 2–1 reviews the equatorial coordinate system.

which they are applicable. Those in use today have been drawn for the epoch 2000.0. In order to point a telescope accurately at any object in the sky whose right ascension and declination are known from this atlas, corrections have to be made night by night to allow for this slow drift. This adjustment is known as precessing the coordinates of an object and is always done before pointing the telescope toward the object. In modern telescope systems, this adjustment is carried out automatically by the telescope control system.

In this project, we will display and measure the precessional motion of the northern extension of the spin axis of the Earth, the north celestial pole, or NCP, by compressing time so that we can "observe" the sky over very long periods. We will also examine and measure the motion of the **vernal equinox** across the sky because of precession. The vernal equinox is one of the two points at which the Earth's orbital plane intersects its equatorial plane and is used as the zero point of right ascension. The Sun passes through the vernal equinox on the first day of spring every year, hence its name.

> Figure 2–15 of Freedman, Geller, and Kaufmann, *Universe*, 9th Ed. illustrates the vernal equinox.

In the final section of the project, we will observe and measure a tiny additional wobble of the spin axis of Earth, known as nutation, which is superimposed on the smooth and long-term precession.

A. Observation of Precessional Motion

You will use *Starry Night Enthusiast*™ to view the northern polar region of the sky, and follow the motion of the north celestial pole (NCP) against the background stars as time is advanced rapidly. The NCP is easily found at the present epoch because it is close to a bright "marker star," Polaris, the so-called Pole Star of the northern hemisphere.

> 1. Launch *Starry Night Enthusiast*™.
>
> 2. Select **Favourites > Observing Projects > Precession and Nutation > Precession.**

The view shows the northern sky as seen from Toronto, Canada, on March 21, 2007, at local midnight. The north celestial pole is labeled and marked by a red cross. Polaris, also labeled, is visible just below the north celestial pole. This wide-angle view is centered on the north celestial pole, the position in space to which the Earth's rotation axis points. The gaze direction is therefore parallel to the rotation axis of Earth. The stars will appear to move as time changes because the view is locked on to the NCP. In reality, the stars represent a fixed background against which the NCP moves.

> 3. With the **Time Flow Rate** set to **50 years, Run time forward** and **backward** to follow the track of the north celestial pole across the sky between the **Date** limits in *Starry Night Enthusiast*™, **4713** BC to **9999** AD.

You will see that, as a result of precession, the north celestial pole appears to move in a circle around a fixed position among the stars. This fixed position is known as the **north ecliptic pole.**

> 4. Select **Favourites > Observing Projects > Precession and Nutation > View Along Axis.** This view is from a location 0.001 AU over the south pole of the Earth, with the view centered on the Earth. Thus, you are looking along the spin axis towards the north celestial pole, hidden behind the center of the Earth in this view. The line from the Earth's center to the north ecliptic pole is perpendicular to the ecliptic plane and is the axis of the cone through which the rotation axis of Earth moves under precession. The cone angle is about 23^1/$_2$°. **Zoom In** to a field of view about **65° × 45°** and use the **Angular Measurement Tool** to measure the angular separation between the center of the Earth and the north ecliptic pole. Since you are looking along the spin axis of Earth, this is equivalent to measuring the separation between the NCP and the north ecliptic pole.

Question 1. What is the angular separation between the center of the Earth and the north ecliptic pole?

5. Select **File > Revert** and with the **Time Flow Rate** set to **50 years**, **Run time forward** and **backward** to follow the track of the north celestial pole across the sky between the **Date** limits, **4713** BC to **9999** AD.

As you **Run time backward** and forward through the range of dates available with *Starry Night Enthusiast*™, note that the north ecliptic pole marks the fixed position around which the stars appear to rotate. Also note how fortunate we in the northern hemisphere are at the present time in history to have a star as bright as Polaris, our "Pole Star," so close to the north celestial pole.

It is interesting to use this time compression to investigate the position of the north celestial pole in history and in the future and answer the following series of questions.

6. Select **Favourites > Observing Projects > Precession and Nutation > Earth Centre.**

This view is nearly identical to that of the previous sequence except that the observing location is at the center of a transparent Earth. The gaze remains along the rotation axis of Earth toward the north celestial pole.

7. Manipulate the **Zoom** and **Time** controls in the toolbar (including adjusting the **Time Flow Rate**) and use the **Angular Measurement Tool** to help you to answer the following questions. (*Hint:* You might want to **Centre** and lock onto the star in question to make some of these measurements.)

Question 2. When is the north celestial pole closest to Polaris, our Pole Star?

Question 3. How close in angle will Polaris be to the north celestial pole on this date?

Question 4. What is the year of closest approach of the north celestial pole to the star Thuban, and what is the angular spacing between north celestial pole and Thuban on this date?

Question 5. To which of these two stars, Polaris or Thuban, did (or will) the NCP approach closest in this precessional motion?

8. Select **File > Revert** and then set the **Date** in the toolbar to **January 1, 4713** BC.

9. Select **View > Constellations > Boundaries**, and **View > Constellations > Labels** from the menu.

10. Use the **Time Flow** controls in the toolbar to determine the sequence of constellations through which the north celestial pole passes in the time period between 4713 BC and 9999 AD.

Question 6. What is the sequence of constellations through which the north celestial pole passes during the period from 4713 BC to 9999 AD.

Question 7. In which constellation is the north ecliptic pole?

Question 8. Does the plane of the Earth's orbit around the Sun shift with respect to the stars over time? How can you tell?

The maximum date available in *Starry Night Enthusiast*™ does not allow one to see the close approach of the NCP to the very bright star Vega in the constellation Lyra, but one can see in 9999 AD that it is moving toward this star.

11. Since you know that the radius of the circle that the NCP makes around the north ecliptic pole is about 23.45°, you can use the **Angular Measurement Tool** to estimate how close the NCP will approach Vega at some time in the future by measuring the angle between Vega and the ecliptic pole.

Question 9. How close will the NCP approach Vega during its precessional motion, at some time in the future?

B. Measurement of Precessional Motion of the North Celestial Pole

We can measure the precessional motion of the rotation axis of Earth easily by finding a time, in history or in the future, when this axis is pointing toward a star in our sky. We can then use this star as a reference point and measure the angle moved by the end of the axis, the north celestial pole, after a specified period of time. An example of this type of close approach occurred in February 1417 AD when the north celestial pole was close to the star designated TYC4641-683-1.

12. Select **Favourites > Observing Projects > Precession and Nutation > TYC4641-683-1** and adjust the field of view to about 20 arcminutes wide. Use the hand tool or scroll bars to adjust the view so that the North Celestial Pole is near the center of the view and the red line passing through the North Celestial Pole is horizontal.

This view shows the north celestial pole and the labeled reference star TYC4641-683-1. (The time step is set to an interval of 19 years for reasons that will become clear in Section D, below.) A time step forward or backward by 19 years will move this reference star from the center toward the edge of the screen. By taking the star's position of closest approach to the north celestial pole as a reference position, you can measure the angular separation between the star and its reference position just above the NCP on these two dates, 19 years on either side of 1417 AD, and then average these values to provide the angular movement of the north celestial pole across our sky in 19 years. From this observation, you can then determine the change in the orientation of the Earth's spin axis per year and calculate the time taken for the axis to complete one full cycle of 360°.

13. **Step time backward** by one 19-year step and use the **Angular Measurement Tool** to measure the angular separation between the reference star and its initial position just above the north celestial pole. Note this value in arcminutes and arcseconds in the column labeled "Measured Separation Between NCP and Star" in the row for 1398 AD in Data Table 1 below.

14. **Step time forward** by two 19-year steps to move the north celestial pole to the opposite side of the reference star and repeat the angular separation measurement, entering this value in Data Table 1 in the row for 1436 AD.

15. Convert your measurements to arcseconds by multiplying the number of arcminutes by 60 and then adding the number of arcseconds to give the total measured separation in arcseconds.

16. Average the results of the previous step to obtain D, the average angular rate of motion of the north celestial pole in a 19-year interval.

17. Divide the average angular rate of motion of the north celestial pole in a 19-year interval by 19 to produce the angular rate of motion of the north celestial pole in arcseconds per year.

18. Divide your result from the previous step by 3600 (the number of arcseconds in one degree) to obtain the angular rate of motion of the north celestial pole, Ω, in degrees per year.

19. Follow the calculations outlined below to determine the revolutionary period of the precessional axis around a full circle.

Data Table 1. Angular Rate of Motion of north celestial pole (NCP) Due to Precession

Date	Measured Separation Between NCP and Star		Measured Separation in Arcseconds	Average 19-year Rate in Arcseconds	Rate per Year (in Arcseconds per Year)	Angular Rate of Motion of NCP
	'	''	('')	('')	('' / yr)	(°/yr)
1398 AD				D	D/19	Ω
1436 AD						

Calculations

The circle followed by the NCP in the sky will have a circumference of $2\pi R$, where $R = 23.45°$ in angular units. Thus, if T, in years, is the time taken for the axis to precess through one full cycle, then the ratio of this time to the 1-year period relevant to the previous measurement equals the ratio of this circumference to the measured annual motion of the NCP, Ω.

Thus,

$$T = (2\pi R/\Omega) = 147.32/\Omega$$

Question 10. How fast does the north celestial pole move across our sky in arcseconds per year?

Question 11. How long will it take the spin axis of Earth to move once around the precessional cone?

C. Motion of the Vernal Equinox

Another manifestation of the precessional motion of the sky is the year-by-year movement of the vernal equinox, one of the points of intersection of the **ecliptic** and the **celestial equator**. The vernal equinox defines the zero point of right ascension in the equatorial coordinate system. This position is also involved in the questionable practice of astrology, though practitioners of astrology do not follow the modern sky and its positions. Nevertheless, two references to the position of the vernal equinox, the first historic and the second modern, have achieved a certain significance. The vernal equinox is sometimes referred to by its historical name of the First Point of Aries since, at a certain point in history, it was found to be within this constellation. We can determine the date when this label ceased to be relevant by tracing the era in which the equinox actually left the constellation of Aries. A more modern reference to the motion of the vernal equinox has entered the lexicon of modern English, namely the "Age of Aquarius," referring to the future. You can use *Starry Night Enthusiast*™ to predict the dawning of this age of infinite promise, which is presumed to start when the vernal equinox enters the constellation of Aquarius.

20. Select **Favourites > Observing Projects > Precession and Nutation > Vernal Equinox.**

You will note that, on the first day of spring in 2005 AD, the Sun is at the vernal equinox as expected. During the year, the Sun moves along the ecliptic at about 1° per year, to return to the vernal equinox each year. *Starry Night Enthusiast*™ follows our present calendar in defining the year as 365 days, compensating for the extra fraction of a day by which a year exceeds this length by adding an extra day every 4 years in **leap years.** Thus the Sun will not return precisely to the vernal equinox unless the time interval is the **tropical year,** a time period defined to ensure this alignment. A slightly longer period, the **sidereal year,** is defined as the time required for the Sun to return to a fixed position with respect to the stars. Notwithstanding these different time periods, you can still observe and measure the drift of the vernal equinox against the background stars because of precession.

21. With the **Time Flow Rate** set at **20 years, Run time forward** and observe the motion of the vernal equinox against the background stars. In this rapid advancement of time, you will note that planets move very rapidly across the sky and that the Sun drifts along the ecliptic because of the time discrepancy discussed above.

You can now use a similar view to determine the two interesting dates referred to above, the date when the vernal equinox left Aries and the date when it will enter Aquarius to begin the "Age of Aquarius."

22. Select **Favourites > Observing Projects > Precession and Nutation > Age of Aquarius.** To reduce confusion, the view does not show the Sun and planets.
23. Manipulate the **Time** controls, **Time Flow Rate** and **Zoom** controls in the toolbar to answer the following questions.

Question 12. In what constellation is the vernal equinox located at the present time?

Question 13. When did the vernal equinox, also called the First Point of Aries, actually leave the constellation of Aries?

Question 14. When will we reach the "Age of Aquarius," defined by the vernal equinox moving into the constellation of Aquarius?

You will note that we are going to have to wait a significant time before we reach this age of great promise!

Since the vernal equinox is also the zero point of the celestial coordinate system, this motion changes the right ascension and declination coordinates of stars and other objects on a day-by-day basis. We can follow and measure this motion.

24. Select **Favourites > Observing Projects > Precession and Nutation > HIP112813.**

The view shows the star HIP112813 centered in the view. Notice that this star lies almost on the ecliptic, the plane of the Earth's orbit around the Sun represented by a green line.

25. Open the **Info** pane and expand the **Position in Sky** layer and note the **RA(Jnow)** and **Dec(Jnow),** the coordinates of the star HIP112813 on this date in May 3319 AD.
26. Position the cursor over the labeled star HIP112813 and measure the angular separation between this star and the vernal equinox.
27. With the **Time Flow Rate** set at **19 years, Step time forward** by one step and notice that now the vernal equinox is very close to the reference star. Also notice from the **Info** pane that the RA and Dec of this star have changed.
28. Watch the RA and Dec of the star in the **Info** pane and **Step time forward** by one more step of 19 years. Then repeat the measurement of the angular separation between the reference star and the vernal equinox.
29. Calculate the average of the two measurements you made of the distance between the vernal equinox and HIP112813 and divide this average by 19 to obtain the rate of motion of the vernal equinox across our sky in arcseconds per year.

Question 15. How fast does the vernal equinox travel across our sky in arcseconds per year?

D. Nutation

In addition to the slow but significant motion of the spin axis of Earth and the vernal equinox because of precession, the varying effects of the gravitational forces from the Moon and the Sun cause a much smaller wobble in the Earth's axis with a period reflecting the motions of the Moon. This wobble is known as nutation and can be observed and measured with *Starry Night Enthusiast™*.

> 30. Select **Favourites > Observing Projects > Precession and Nutation > Nutation.**

In this 1° field of view, a reference star, TYC4662-45-1, lies close to the north celestial pole. In this view, the north celestial pole will remain fixed as time advances. Thus, although the north celestial pole, which is defined by the rotation axis of the Earth, is in reality moving across the sky because of precession, it will appear to remain stationary while the stars will appear to move. Running time forward, you will see the precessional motion of the pole reflected in horizontal motion of the stars.

> 31. With the **Time Flow Rate** set at **1 year**, **Run time forward** and observe the motion of the stars.

You will note that, superimposed on the horizontal precessional motion of the north celestial pole shown by the horizontal motion of the stars, there is a small but perceptible wobble in the vertical positions of these stars as they traverse the sky.

> 32. Select **File > Revert** and replay this sequence to show the effect of this nutational motion of the Earth's spin axis on the effective motions of the stars in our chosen reference frame.

In the next sequence of observations, you will measure this nutational motion and quantify it.

> 33. Select **File > Revert.**
> 34. Change the **Date** in the toolbar to **June 1, 2074** AD.
> 35. Change the **Time Flow Rate** to **2 years.**
> 36. **Zoom In** to a field of view about **18'** wide.
> 37. Use the **Hand Tool** to drag the view so that the north celestial pole is near the center of the screen and the red line across the view that represents the celestial meridian is horizontal. The reference star TYC4662-45-1 should be visible near the left edge of the view.
> 38. Use the **Angular Measurement Tool** to measure the perpendicular distance from the star to the reference meridian (horizontal red line). Note the year and your measurement in Data Table 2 below.
> 39. **Step time forward** in 2-year steps and repeat this angular measurement after each step.

Data Table 2. Angular Distance of TYC4662-45-1 from Reference Meridian

Year	Angular Distance in Arcseconds ('')

The data points in Data Table 2 can be plotted upon a sheet of graph paper to reveal the pattern of the nutation motion of the north celestial pole. Plot the angular distance of the star TYC4662-45-1 from the reference meridian along the vertical axis of the graph and the date along the horizontal axis. From the graph you can determine the relevant parameters of this motion: its period and its amplitude.

Alternatively, you may wish to enter the values into a spreadsheet. The following steps guide you in using Excel to plot your data.

40. Enter the year in column A of the spreadsheet and the angular distance of the star to the reference line in column B. Highlight all of the data points and select **Insert > Chart...** from the menu.

41. In Step 1 of the Chart Wizard select the **XY (Scatter)** chart type and the subtype that shows both data points and a line connecting them. Then click **Next**.

42. In Step 2 of the Chart Wizard, Excel will have automatically determined that the data series is in columns. Click **Next**.

43. In Step 3 of the Chart Wizard, you should give the chart a **Title**, such as "Nutational Motion of NCP" and also provide a title for the **Value (X) axis** as **Year** and the **Value (Y) axis** as **Angular Distance from Meridian ('')**. Under the **Gridlines** tab, choose **Major gridlines** for both the x- and y-axis. Under the **Legend** tab, you can turn the option to **Show legend** off and then click the **Finish** button.

Question 16. Which of the following options best describes the line connecting the data points?

    a) A straight horizontal line

    b) A straight line sloping up over time

    c) A straight line sloping down

    d) A sinusoidal line (i.e., a wavy line that goes alternately up and down over time)

Question 17. If the star shows a periodic variation, what is the period of this variation (i.e., how many years does it take for the graph to move from a maximum value through a minimum value and back to a maximum value)?

Question 18. What is the amplitude of this variation, if any? (Amplitude is the difference between the maximum value and the minimum value divided by two; i.e., one-half of the full variation of the angular separation of the star from the reference line.)

E. Conclusions

The relatively smooth precessional motion of the Earth's spin axis, measured in Sections A, B, and C, is caused by the average effect of the differential gravitational forces from the Moon and the Sun acting on the Earth's equatorial bulge. These forces do not always act in the same direction, however, because of the motion of the Moon along an orbital path that is inclined to the ecliptic plane. The points where the Moon's orbital path crosses the ecliptic, the nodes of the Moon's orbit, slide around the ecliptic plane with a period of about 18.6 years. As seen in Section D, this produces small differences in the gravitational influence of the Moon on the Earth's bulge, resulting in the small nutational wobble superimposed on the smooth precessional coning of the Earth's spin axis.

The Moon's Motions and Phases

10

The Moon orbits the Earth in 27.3 days with respect to the background stars. This period is known as a **sidereal month.** The time period for the Moon to return to the same position with respect to the Sun in its orbital motion, the **synodic month,** is 29.5 days. The synodic month is longer than the sidereal month because in the time required for the Moon to complete an orbit around the Earth with respect to the stars, the Earth-Moon system has moved some distance along its orbit around the Sun. As seen from Earth, the Sun has moved against the background stars. Thus, the Moon must move a corresponding distance to be in the same direction in the sky with respect to the Sun once more.

> The sidereal and synodic months are discussed in Section 3–2 and illustrated in Figure 3–5 of Freedman, Geller, and Kaufmann, *Universe*, 9th Ed.

The Moon shows its complete cycle of phases in the course of a synodic month. This arises from the fact that the Moon is a spherical body illuminated by the Sun. One half of the Moon's surface is in light and the other half is in shadow at any time. The proportion of the illuminated hemisphere visible from Earth depends on the position of the Moon with respect to the orientation of the Earth and Sun.

> This cycle of lunar phases is described in Section 3–1 and Box 3–1 and illustrated in Figure 3–2 of Freedman, Geller, and Kaufmann, *Universe*, 9th Ed.

In the monthly cycle of phases, the Moon appears first as a very thin crescent in the western sky, quite close to the Sun. It then proceeds to show more of its illuminated side to Earth as it moves farther from the Sun in the sky. In this growing half of the cycle, or **waxing** phase, it will pass through **crescent, first quarter,** and **gibbous** phases before reaching **full Moon.** At its **full** phase, the Moon is opposite to the Sun in our sky and the entire sunlit hemisphere of the Moon is visible from the Earth. Its angular separation from the Sun then begins to decrease as it moves into the diminishing half of the cycle, or its **waning** phase. It proceeds through **gibbous, third quarter,** and **crescent** phases until it is again in the same direction in the sky as the Sun and its dark hemisphere faces the Earth at **new Moon.**

In this project, you will observe the motion and phases of the Moon and investigate this relationship between the Moon's phases and its position with respect to the Sun in the sky.

A. The Moon's Motion Across the Sky and its Angular Speed

1. Launch *Starry Night Enthusiast*™ and open the **Find** pane. Click the blue menu button to the left of the label for the **Moon.** In the menu that pops up, ensure that the **Enlarge Moon Size** option is checked. (Note: In this mode, the Moon will appear larger than it's natural size for wide-angle views.)

2. Open the **Preferences** dialog (Windows users select **File > Preferences** and Macintosh users select **Starry Night Enthusiast > Preferences**) and under the **Cursor Tracking (HUD)** options check the **Name** and **Object Type** in the **Show** list.

3. Select **Observing Projects > Moon's Motion > Moon's Angular Speed** from the **Favourites** menu or pane.

The view shows the sky looking south from Chicago at 12:45:00 AM Central Daylight Time on July 28, 1999. The full moon is visible in the center of the view.

4. Change the **Time Flow Rate** to 300 × and observe the motion of the Moon and stars in the night sky.

With time flowing at 300 times its normal rate, the westward drift of the sky resulting from Earth's eastward rotation becomes obvious. Our view of the sky from the rotating platform of Earth shows the stars and the Moon moving together in this westward direction.

The Moon shows an additional motion compared to the stars. If you watch carefully as the sky turns westward, you will see that the Moon appears to drift against the background stars. The following sequence allows you to observe this additional motion of the Moon more clearly.

5. Select **File > Revert** from the menu.

6. Open the **Find** pane and type **Sigma Capricorni** in the edit box at the top of the pane followed by the **Enter** key to center the gaze upon this star.

7. **Zoom** In to a field of view of about **15° × 10°**.

8. Set the **Time Flow Rate to 300 ×** and observe the motion of the Moon and the stars. With the gaze centered on the star Sigma Capricorni, the sky will not now move westward, but the horizon will move eastward. Observe the motion of the Moon with respect to the background stars until Sigma Capricorni has set beneath the western horizon. To repeat the animation, select **Edit > Undo Time Step** and then **Edit > Redo Time Step** from the menu or simply reverse the direction of time flow.

Question 1. Is the Moon east or west of the star Sigma Capricorni when the animation begins?

Question 2. Toward which direction is the Moon from the star Sigma Capricorni when this star sets?

Question 3. In which direction does the Moon appear to drift against the background stars as time flows forward?

Question 4. This independent motion of the Moon against the diurnal motion of the much more distant stars results from the Moon's orbital motion around the Earth. From your observations, in which direction does the Moon orbit the Earth? (Hint: Sketch the expected view of Earth and Moon from above the Earth's north pole to help with this question.)

Observing the Moon at successive intervals of 1 sidereal day will allow you to explore this independent motion of the Moon more precisely since the stars will return to the same position in the sky every sidereal day. The independent motion of the Moon will show up as a change in its position against the background stars.

9. Select **File > Revert** from the menu. Notice that the **Time Flow Rate** is set to **1 sidereal day.**

10. Click the **Step time forward** button in the toolbar to advance time by 1 sidereal day.

In the interval of 1 sidereal day, the stars have returned to the same position in the sky but the Moon, having proceeded some distance along its orbit around the Earth, has moved against the background stars. You can measure how far the Moon moves in our sky in 1 sidereal day.

11. Select **File > Revert.** Use the **Angular Measurement Tool** to measure the angular separation between the Moon and the labeled star Albaldah. Note this value along with the date in the Data Table below.

12. **Step time forward** to advance time by 1 sidereal day and repeat the measurement of the angular separation between the Moon and Albaldah. Note the date and angular separation in the Data Table.

13. Calculate the difference between these two angular separations and enter this value in the Data Table. The difference between these measurements is the average angular speed of the Moon over the chosen sidereal day.

Data Table. Angular Separation of the Moon from the Star Albaldah

Date	Angular Separation	Difference in Angular Separation

Question 5. What is the angular speed of the Moon against the background stars, in degrees per sidereal day?

Question 6. Based on your previous answer, approximately how long would it take for the Moon to return to the same position relative to the stars?

The Earth's orbital motion around the Sun produces an apparent motion of the Sun against the background stars when viewed from Earth. In the course of a year, the Sun traces a complete path around the sky called the **ecliptic**. The ecliptic is the intersection of the plane of the Earth's orbit around the Sun with the celestial sphere. In the next sequence, we will track the Moon's motion with respect to the ecliptic.

> The ecliptic is described in Section 2-5 of Freedman, Geller, and Kaufmann, *Universe,* 9th Ed.

14. Select **File > Revert** from the menu.

15. Use the **Step time backward** button in the toolbar to move backward in time a total of 4 sidereal days to **1:00:44** AM Daylight time on **July 24, 1999.**

16. Select **View > The Ecliptic** from the menu to display the ecliptic on the sky.

17. **Step time forward** again using steps of 1 sidereal day and watch the Moon's motion with respect to the ecliptic.

The Moon's orbit is inclined at about 5° to the ecliptic plane. Thus, the Moon is seen to follow a path across the sky that is close to but not on the ecliptic. The Moon's path crosses the ecliptic plane at two points during each orbit around the Earth.

These points are called the nodes of the Moon's orbit. The node at which the Moon crosses the ecliptic from south to north is called the **ascending node,** while that at which the Moon crosses the ecliptic from north to south is called the **descending node.** The nodes are significant because it is only when both the Sun and the Moon are near a node that a **solar** or a **lunar eclipse** can occur. Indeed, that is how the **ecliptic** got its name.

> The nodes of the Moon's orbit are discussed in Section 3–3 and illustrated in Figure 3–6 of Freedman, Geller, and Kaufmann, *Universe*, 9th Ed.

Question 7. On what date during the observing period in the previous sequence does the Moon cross the ecliptic?

Question 8. Is this the ascending or the descending node of the Moon's orbit?

B. The Waxing Phases of the Moon

To demonstrate the phases of the Moon, we must observe the sky at different times of the night on different days. The best time to observe the waxing phases is just after sunset when the Moon appears in the western sky. The phases around full Moon are best observed near midnight, while the waning phases can be seen most effectively in the dawn sky.

18. Select **Observing Projects > Moon's Motion > Waxing Moon** from the **Favourites** menu or pane.

In this view of the sky looking west from Chicago at 8:20:00 PM, Central Daylight Time, on July 13, 1999, the Sun is labeled and is near to the west-northwest horizon and is about to set. The Moon, also labeled, is barely visible as a faint silver disk in the twilight. The portion of the Moon that is not lit by the Sun is visible from Earth at this time because of **earthshine.** This faint glow is caused by sunlight that has been reflected from the surface and clouds of the Earth to illuminate the Moon. This appearance of the thin crescent Moon was referred to in earlier times as "the old moon in the new moon's arms." Note that the Moon is in the same general direction in the sky as the Sun at this time, just after new moon, and that the Moon looks larger than the Sun because the **Enlarge Moon Size** setting is selected.

19. Click the **Sunset** button in the toolbar to set the time to local sunset.

20. **Step time forward** two days to **July 15, 1999.** Note that each time step is 1 solar day rather than 1 sidereal day. Thus, the local time remains the same, close to sunset.

21. Position the **Hand Tool** over the bright object just below the Moon and use the **HUD** to identify this object.

Question 9. What object appears near to the Moon on July 15, 1999?

Question 10. In which way do the "horns," the sharp endpoints of the crescent shape of the illuminated portion of the Moon, point with respect to the direction of the Sun?

Question 11. Toward which compass direction in your sky do the "horns" of the waxing crescent Moon point, as seen from the Earth?

If you were to continue to step forward in time, the Moon would eventually disappear off the left side of the screen. To watch the Moon go through its phases, you will need to lock the view onto the Moon so that it remains centered on the screen as you manipulate the time flow.

22. **Centre** the view on to the Moon.
23. Select **View > Hide Daylight** from the menu.
24. **Step time forward** by 1 day.

Now, the Moon remains stationary in the center of the view, while the sky *and the horizon* shift to the right (west). By keeping the Moon centered in the field of view, the direction of your gaze shifts eastward, night by night.

25. **Step time forward** to **July 20, 1999**, and watch the phase of the Moon change from day to day.

As time progresses and the Moon moves away from the Sun in the sky, the proportion of the Moon's illuminated hemisphere that can be seen from the Earth increases or waxes. On July 13, the Moon was a very slender crescent close to the Sun. By July 20, the Moon's orbital motion has carried it some distance away from the Sun in the sky and it shows the **first**, or waxing, **quarter phase.**

You can measure the angular distance between the Sun and the Moon as seen from Earth and thereby gain some insight into how the relative positions of the Sun, Moon, and Earth produce the various phases of the Moon.

26. Use the **Hand Tool** to drag the sky to the left until both the Moon and the Sun are visible in the view.
27. Measure the angular separation between the Sun and the Moon when the Moon is at first quarter phase.

Question 12. What is the angular separation between the Moon and the Sun as seen from Earth when the Moon is at first quarter phase?

Question 13. As seen from Earth, which side of the Moon's disk, east or west, is illuminated by the Sun when the Moon is at first quarter phase?

28. **Centre** the view on the **Moon** once more.
29. **Step time forward** in 1-day intervals to **July 27, 1999**, and observe the successive waxing gibbous phases of the Moon.

By July 27, the Moon is almost full. Notice that the Moon is close to the ESE point on the horizon, or 180° away from the Sun's position at WNW. For the Moon to be fully illuminated as seen from the Earth, it must be in the opposite part of the sky from the Sun.

C. The Waning Phases of the Moon

30. **Step time forward** 1 day to **July 28, 1999.**

The Moon is just below the horizon at sunset on July 28, 1999. This opposition of Sun and Moon in the sky is a condition for full moon. Because of the eastward motion of the Moon in its orbit around the Earth, it rises later each day. Clearly, you will need to change the time at which you make your observations during the waning phase to ensure that the Moon is above the horizon.

31. Change the **Time** in the toolbar to **01:00:00** AM on **July 28, 1999**.

32. Watch the waning gibbous phase of the Moon as you **Step time forward** in intervals of 1 day to **01:00:00** AM on **August 4, 1999**. Identify the bright object that is visible just above Moon at this time.

Question 14. Which side of the Moon's disk is illuminated by the Sun as seen from the Earth when the Moon is at third quarter?

Question 15. In which direction, relative to the Sun does the Moon move in the waning phases?

Question 16. Without using the Hand Tool to measure it, what do you expect to be the angular separation of the Moon and Sun as seen from the Earth when the Moon is at third quarter?

Question 17. To which object does the Moon pass close on August 4?

33. **Step time forward** 1 further day to **01:00:00** AM on **August 5, 1999**. Identify the bright object above the Moon at this time.

Question 18. To which object does the Moon pass close on August 5?

34. To continue to follow the waning phases of the Moon, change the **Time** in the toolbar to **5:45:00** AM on **August 5, 1999**.

35. **Step time forward** in 1-day intervals to **5:45:00** AM on **August 9, 1999**, and observe the phase of the Moon wane from third quarter to crescent.

Question 19. In which way do the "horns" of the waning crescent moon point with respect to the direction of the Sun?

Question 20. Toward which compass direction do the "horns" of the waning crescent moon point as seen from the Earth?

Question 21. Compare your answers to Questions 10 and 11 with your answers to Questions 19 and 20.

36. **Step time forward** 1 solar day to **5:45:00** AM on **August 10, 1999**.

By 5:45:00 AM on August 10, the Moon is near the eastern horizon just before sunrise. Thus, since July 13, it has nearly completed the entire cycle of phases. By August 12, 1999, the Moon will once again be a thin crescent in the western sky at sunset.

37. **Step time forward** to **August 12, 1999**, and then click the **Sunset** button in the toolbar.

D. Positions of the Sun, Moon, and Earth and the Moon's Phases

In this section of the project, you will observe the phases of the Moon from a unique point of view that will allow you to see how the relative positions of Earth, Moon, and Sun lead to the Moon's cycle of phases as these objects move through their respective orbits.

38. Select **Observing Projects > Moon's Motion > Relative Positions of Sun, Moon and Earth** from the **Favourites** menu.

The view is of the Sun and Moon from a location above the north pole of the Earth on September 17, 2001. The Moon, labeled but virtually invisible, is very close to the Sun in the sky. Since the Moon is between the Sun and the Earth, the side of the moon facing Earth is also facing away from the Sun and, as a consequence, is dark. This is the **new phase** of the Moon. The Moon and the Sun are in the same direction in the sky, as seen from the Earth.

39. Use the **Hand Tool** to drag the sky upward until the Earth comes into view on the bottom of the screen to understand the orientation of your viewing location with respect to the Earth, Moon, and Sun.

Question 22. Considering the fact that the distance from the Earth to the Sun is roughly 400 times the distance from the Earth to the Moon, what geometric term best describes the relative position of the Sun, Earth, and Moon when the Moon is in the new phase (e.g., right triangle, straight line, etc.)?

40. Select **File > Revert** to center the Moon in the view once again.
41. **Step time forward** in steps of 1 day until the Moon reaches first quarter phase. You will note that the view is locked onto the Moon.

You will see that, as the Moon moves away from the Sun direction, a sliver of reflected sunlight begins to be visible on the western edge, or limb, of the Moon as seen from the Earth. Stepping forward day by day reveals progressively more and more of the sunlit side of the Moon. This is the waxing phase of the lunar cycle. As you can see, between 3 and 4 days after new moon, the Moon is in the first or waxing crescent phase and 7 days after new Moon, it reaches the first quarter phase. The relative positions of Moon and Sun can be seen by using the **Hand Tool** to drag the view toward the left until the Sun becomes visible at the right (west) edge of the view window. Since the screen encompasses a field of view of 100°, the Sun and the Moon are about 90° apart in the sky as seen from the Earth.

Question 23. What geometric term best describes the relative positions of the Sun, Moon, and Earth at first quarter phase of the moon?

42. **Centre** the view on the **Moon** once again.
43. Continue to **Step time forward** in 1-day steps and observe the Moon through the waxing gibbous phase to the point where the entire surface of the Moon that faces the Earth is now lit by the Sun, on **October 2, 2001**.

This is **full moon,** where again the Earth, Sun, and Moon lie roughly on a straight line, but now the Earth is between the Sun and the Moon.

44. With the Moon centered in the view, note the azimuth of the **Gaze** direction in the toolbar.

45. Use the **Hand Tool** to drag the sky to the left until the Sun appears in the view. **Centre** the view on the Sun and note the azimuth of the **Gaze** direction in the toolbar.

Question 24. From the difference in azimuth of the Sun and the Moon that you observed in the previous sequence, what is the angular separation between the Sun and the Moon when the Moon is in its full phase?

46. **Centre** the view on to the **Moon** once more.

47. Continue to **Step time forward** in intervals of 1 day to observe the Moon through waning gibbous, last quarter, and waning crescent phases to new moon again on **October 16, 2001.**

E. The Synodic Month

48. Select **Observing Projects > Moon's Motion > Synodic Month** from the **Favourites** menu or pane.

The view is from Chicago on July 12, 1999. The horizon and daylight have been hidden to give an unobstructed view of the Moon. The new moon appears near to the Sun in the sky. Note the background of stars that surround the Moon, particularly the position of the bright star **Mekbuda,** which is labeled.

The time required for the Moon to complete a full cycle of phases from new moon to new moon is called the **synodic month.** Since the phases of the Moon depend on its relative position with respect to the Sun as seen from Earth, the synodic month is also the time required for the Moon to make one complete orbit with respect to the Sun, as viewed from Earth.

49. **Step time forward** in intervals of 1 day. Stop after 27 such daily steps and note the position of the Moon relative to the reference star Mekbuda. A little more than 27 days, or 1 sidereal month, is required to bring the Moon back to the same region of the sky with respect to the background stars.

50. **Step time forward** by two more steps of 1 day to bring the Moon back to its new phase. You will see that about 29.5 days, or 1 synodic month, would have brought the Moon back to the same position with respect to the Sun.

51. Select **File > Revert** and repeat the sequence to answer the following questions.

Question 25. How many days elapse between a) new moon and first quarter? b) new moon and full moon? c) new moon and the next new moon?

Question 26. How long, in days, is a synodic month?

52. Select **File > Revert.**

53. Select **View > The Ecliptic.**

54. Position the **Hand Tool** over the Moon and right-click (Macintosh users Ctrl-click) the mouse button to display the **Object Contextual Menu.** Select **Orbit** to show the orbit of the Moon in the view.

55. Make sure that the view is centered on the **Moon**.

56. **Step time forward** in 1-day steps to **August 10, 1999,** to observe the Moon's path relative to the ecliptic and answer the following questions.

Question 27. How often does the Moon cross the ecliptic plane in 1 synodic month?

Question 28. On what date does the Moon pass through the ascending node of its orbit?

Question 29. On what date does the Moon pass through the descending node of its orbit?

F. The Phases of the Earth

In the next sequences, you will make observations of the Earth from the Moon and explore the phases that the Earth displays to an observer on the Moon over the same time period as the previous sequence.

57. Select **Favourites > Observing Projects > Moon's Motion > Earth's Phases** from the menu.

The view is of the Earth from the surface of the Moon, at a location near a crater named Birmingham. To see where this is on the Moon, you can open the **Viewing Location** dialog box by selecting **Options > Viewing Location...** from the menu and clicking on the **Map** tab in the dialog box. A red X on the map marks the current viewing location. Click the **Cancel** button to close the dialog box and return to the view.

The date is the same as that at the beginning of the last section, July 12, 1999, when the Moon was at new phase as seen from Earth. Now, the view is of the Earth as seen from the Moon.

Question 30. At this time, when the Moon as seen from Earth is in its new phase, at what phase is the Earth as seen from the Moon?

58. **Step time forward** in intervals of 1 day to the date of full Moon as seen from Earth, namely **July 27, 1999,** and observe the changing phase of the Earth as seen from the Moon.

Question 31. Is the Earth waxing or waning?

Question 32. What is the phase of the Earth as seen from the Moon on the date at which the Moon is full as seen from Earth (July 27, 1999)?

Question 33. On July 27, 1999, in which direction in the sky with respect to the Sun is the Earth, as seen from the Moon?

59. **Step time forward** in 1 day steps to the date of the next new Moon (**August 11, 1999**).

Question 34. In this period of time, while the Moon wanes as seen from Earth, what phases does the Earth go through as seen from the Moon?

It is interesting to use *Starry Night Enthusiast*™ to observe the Earth as seen by astronauts on the Moon.

60. Select **File > Revert** from the menu.

61. Change the **Time Flow Rate** to **1 hour**.

62. **Run time forward** to watch the horizon and the Earth as time progresses through at least one full cycle of phases of the Earth.

This provides a fascinating view of our Earth rotating on its axis as it slides across the lunar sky, showing phases equivalent to those of the Moon seen from Earth.

Another glimpse of life for an astronaut on the Moon can be seen as you watch the abrupt arrival of sunrise on the Moon. Since there is no atmosphere, there will be no twilight, so the Sun suddenly floods the landscape with light.

Question 35. Notice that the horizon does not get in the way at any time during this sequence. Can you explain why?

G. Lunar Occultation of a Star

One striking consequence of the lunar motion is that, as the Moon slides eastward against the background stars, it occasionally covers and uncovers stars in its path. Such an event is called an **occultation**. You met this phenomenon briefly in Section A when you observed the motion of the Moon against a reference star Sigma Capricorni. An occultation of a star by the Moon is an exciting event to observe through a telescope. When the Moon covers a star, the event is called a disappearance, and when it uncovers a star, it is known as a reappearance event. These events are particularly exciting to watch when they occur against the unlit limb (edge) of the Moon. Because the stars are (almost) true points of light in the sky, and because the Moon has no atmosphere, the star vanishes almost instantaneously from view in a disappearance event against the dark limb of the Moon. In fact, a star's angular size can sometimes be determined by an accurate observation of the disappearance of the star with a high-speed photometer during an occultation. The following sequence simulates these occultation events.

63. Select **Observing Projects > Moon's Motion > Occultation** from the **Favourites** menu or pane. In this view, the gaze is locked on the Moon and the field of view encompasses only 3°.

64. Change the **Time Flow Rate** to 300× and observe the motion of the Moon against the background stars. You will note that several faint stars become occulted by the dark advancing limb of the Moon.

An occultation event is particularly interesting when circumstances are such that only the mountains and crater walls of the Moon's southern or northern polar regions pass in front of the star. This is what is known as a **grazing occultation** and, from the correct location on Earth, the star is seen to wink out several times as it moves behind these higher regions at the Moon's poles. This type of occultation provides an opportunity for observers to make important observations of regions of the Moon that are otherwise difficult to measure. If several observers watch the occultation through telescopes from sites spaced at right angles to the shadow path of the Moon across the Earth and carefully time the disappearances and reappearances of the star, a detailed profile can be constructed of the topography along the Moon's limb.

A similar procedure can be used when an asteroid occults a star. Asteroids are too far away for us to be able to see them as much more than points of light, even through the largest telescopes. However, occultation observers scattered across the shadow path of the asteroid on the Earth see different parts of the asteroid passing in front of the star. Careful timings of the disappearance and reappearance of the star from several different locations can tell us the shape and size of the asteroid. A star has occasionally been seen to disappear and reappear a second time for a particular asteroid, indicating that a small moon accompanies this asteroid.

Question 36. An occultation event can be a disappearance or a reappearance and can occur against the bright limb or the dark limb of the Moon. For each combination below, identify whether the Moon is waxing or waning.

a) Disappearance at the dark limb

b) Reappearance at the bright limb

c) Reappearance at the dark limb

d) Disappearance at the bright limb.

H. Conclusions

In this project, you have observed the phases of the Moon from different viewpoints as the Moon moves around the Earth in its orbit, while noting the relative positions of the Moon and the Earth with respect to the Sun in the sky for each phase. You have measured the Moon's angular speed across our sky and you have compared the sidereal and synodic periods of the Moon as measured from the Earth. You have also examined the phases of the Earth as they would appear to an explorer on the Moon and have seen that these phases are complementary to those of the Moon as seen from Earth.

Solar Eclipses

11

A solar eclipse occurs when the Moon passes in front of the Sun and casts its shadow upon the Earth's surface. Most observers within this shadow on the Earth will see the Moon obscure, or occult, a portion of the solar disc in a partial solar eclipse. By a fortunate coincidence, the angular sizes of the Sun and the Moon are about equal in our sky. Consequently, it is possible for the Moon and Sun to be aligned so that the Moon can completely obscure the Sun when seen from specific locations on the Earth. This marvellous spectacle of nature is a total solar eclipse.

Eclipses are discussed in Section 3–3 of Freedman, Geller, and Kaufmann, *Universe*, 9th Ed. Solar eclipses in particular are discussed in Section 3–5.

The Moon is in the same direction in the sky as the Sun as seen from Earth only when it is at its new phase. If the plane of the Moon's orbit were exactly in the Earth's orbital plane, the ecliptic plane, then a solar eclipse would occur at every new Moon. In fact, the orbital plane of the Moon is inclined to the ecliptic plane by about 5°. Thus, the Moon usually passes north or south of the Sun when it is in its new phase and does not align closely enough to the Sun to produce an eclipse. The Moon's inclined orbit intersects the ecliptic plane at two points called the **nodes** of the Moon's orbit. The node is known as the ascending node when the Moon approaches it from south of the ecliptic plane and the descending node when the Moon passes through this position from north of the ecliptic plane. When a new Moon occurs near to one of these nodes, the Sun and Moon are aligned closely enough that the Moon will eclipse the Sun as seen from some regions of the Earth.

Figure 3–6 in Freedman, Geller, and Kaufmann, *Universe*, 9th Ed., illustrates the inclination of the Moon's orbit to the ecliptic.

Figure 3–7 in Freedman, Geller, and Kaufmann, *Universe*, 9th Ed., illustrates the nodes of the Moon's orbit and the conditions required for a solar eclipse.

The Sun appears to move completely around the ecliptic in one year as seen from the Earth. Therefore, twice a year it is near to a node of the Moon's orbit. While the Sun appears to move against the background stars at only about 1° per day, the Moon moves at about 13° per day in its monthly orbit around the Earth. Consequently, while the Sun moves slowly past a node, it is possible for the relatively fast Moon to move into its new phase and produce an eclipse. The maximum distance that the Sun can be from a node and still be eclipsed by the Moon depends on several factors. These factors include the apparent angular sizes of the Sun and the Moon as seen from Earth and the parallax of the Moon from the distance of the Earth. The interval over which the Sun remains within these ecliptic limits and close enough to a node for an eclipse to occur is known as an **eclipse season**. Each such season lasts for about four weeks and the interval between successive seasons is a little less than six months.

IMPORTANT! Never look directly at the real Sun, even when it is partially eclipsed, without proper eye protection (Welder's glass #14 or specially designed aluminized glass or plastic filters). In the absence of proper eye protection, observe a projected image of the Sun. Looking at the Sun without proper protection can produce serious and permanent eye damage!

The Moon is much smaller than the Sun and therefore the darkest part of the shadow cast by the Moon into space, the umbra, is conical in shape. On average, the length of this umbral cone is slightly less than the average distance between Earth and Moon. Thus, even if the Moon and Sun are aligned in the sky, the umbra may not reach the surface of the Earth. Under these conditions, observers in the path of the Moon's shadow will see the Moon somewhat smaller than the Sun and, at the moment of maximum eclipse, the Moon will appear encircled by a narrow ring of the Sun's disk. This is known as an **annular solar eclipse.** Observers who are outside the central region of the eclipse will see only part of the Sun's disk covered and will therefore be in the **partial eclipse** zone.

There are times when the elliptical orbit of the Moon brings it closer than average to the Earth. If an eclipse occurs during this time, the Moon's umbra reaches the Earth and observers within this narrow shadow region will see a **total solar eclipse** in which the Moon occults the entire solar disc. Just as in an annular eclipse, observers on either side of the path of totality will see a **partial solar eclipse.**

The total eclipse of the Sun is one of nature's finest spectacles. Only at this time can you see the tenuous but hot outer atmosphere of the Sun directly without protection for the eyes or special instrumentation. Indeed, this geometry is such that many phenomena can be seen and measured *only* during this brief eclipse period known as **totality.** The Sun's outer atmosphere directly above the Sun's visible surface is comprised of the **chromosphere** overlaid by a very tenuous but extremely hot outer region called the **corona.** Both have been studied extensively over the past century using eclipse observations.

A. Conditions Required for Solar Eclipses

The conditions for a solar eclipse to occur are that the Moon must be at its new phase and close to the ecliptic plane near to a node in its orbit, as discussed above. You can demonstrate these conditions using the following procedure.

1. Select **Favourites > Observing Projects > Solar Eclipses > Eclipse Requirements.**

This view is from the center of a transparent Earth with the gaze centered and locked on the star *Mebsuta*, in the constellation Gemini. On this date and time, the Moon is new, thus presenting its dark hemisphere toward the Earth. It is not visible but a label indicates its position.

2. If you wish, you can adjust the brightness of the dark side of the Moon so that it is visible in the view. Select **Options > Solar System > Planets-Moons...**from the menu. In the **Planets-Moons Options** dialog box, check the box for the **Show Dark Side** option. Adjust the slide control to the right of this option so that it is at about the midpoint of the brightness scale. This will produce a dim and artificially illuminated image of the dark hemisphere of the Moon. Also check under **Options > Other** in this dialog box that the **Enlarge Moon Size at Large FOVs** is unchecked.

3. **Right-click** (Macintosh users **Ctrl-click**) over the Moon to bring up the object contextual menu and select **Show Info.** In the **Info** pane for the Moon, expand the **Other Data** layer. Note that the **Age of Moon** is New, 0 days old. Leave the **Info** pane open.

As you see, one requirement for a solar eclipse is fulfilled in this view, namely that the Moon is new, 0 days old. However, the Moon is too far from the ecliptic to occult any part of the Sun on this date.

4. Use the **Angular Measurement Tool** to determine the angular separation between the new Moon and the Sun. Note that, when making these measurements, *Starry Night Enthusiast*™ measures the separation between the centers of the Sun and Moon.

Question 1. What is the angular separation between the Sun and the new Moon of June 25, 2006?

Question 2. In which direction is the Moon from the Sun on this date?

5. With the **Time Flow Rate** set at **30,000 ×**, click the **Run time forward** button.

As time advances, the Moon drifts quickly eastward out of the view as it makes its monthly orbit around the Earth. Since the gaze is locked onto a star, the Sun also appears to drift much more slowly eastward against the background stars as a result of the yearly motion of the Earth in its orbit around the Sun.

6. As the Moon reaches the left side of the view, **Stop Time** and adjust the **Date** to move the Moon around its orbit to bring it into view to the right of the Sun. Adjust the hours and minutes of the **Time** in the toolbar until the **Age of Moon** as indicated in the **Info** pane is once again **New, 0.00 days old.**

7. Measure the angular separation between the center of the Sun and the center of the Moon on this date.

Question 3. Is the angular separation between the Sun and the Moon increasing or decreasing?

8. **Centre** the view on to the Sun (open the **Object Contextual Menu** for the **Sun** and select **Centre**).

9. Use the **Time Flow** controls in the toolbar to advance time to the next new Moon on August 23, 2006. To do this easily, **Step time forward** by **1 lunar month** and adjust the **Time** in the toolbar until the **Info** pane indicates that the Moon is **New, 0.00 days old.**

10. Measure the angular separation between the Sun and the new Moon on this date.

In the view, you will notice a wedge-shaped icon at the point where the Moon's orbit crosses the ecliptic to the left of the current positions of the Sun and Moon. This icon indicates a node of the Moon's orbit.

Question 4. Is this node, visible in the present view, the ascending or the descending node?

11. Set the **Time Flow Rate** to **1 lunar month** and **Step time forward** by one step to **September 22, 2006.**

12. **Zoom In** to a field of view about **10°** wide.

Both the Sun and the Moon have passed the node of the Moon's orbit at this time and the Moon is not quite new.

13. Adjust the **hours** and **minutes** of the **Time** in the toolbar to move the Moon to its new phase, 0.00 days old, as before.

14. Measure the angular distance from the Sun to the node of the Moon's orbit.

Question 5. Why does this new Moon partially eclipse the Sun, even though both the Sun and the Moon have moved beyond the node of the Moon's orbit?

Question 6. What is the angular distance from the Sun to the node?

All of the conditions for a solar eclipse are met in this view. The Moon is new, and the Sun is near one of the nodes. Since the Moon's orbit intersects the ecliptic at two nodes, we would expect that the requirements for a solar eclipse would be met again when the Sun reaches the other node in approximately six months.

15. With the view centered on the **Sun, Zoom Out** to a field of view **100°** wide.

16. Set the **Time Flow Rate** to **1 lunar month** and **Step time forward** by six 1-month steps to **March 18, 2007.**

Question 7. Is the Sun closest to the ascending node or the descending node on this date?

17. Adjust the **hours** and **minutes** of **Time** to move the Moon to its New phase, 0.00 days old.

18. Measure the angular distance from the Sun to the node.

19. If necessary, **Centre** the view on the **Sun** and **Zoom In** to a field of view about **10°** wide.

20. Measure the angular distance between the Sun and the Moon.

21. **Right-click** over the Sun and select **Show Info** from the **Object Contextual Menu**. Make a note of the **Angular Size** of the Sun shown in the **Other Data** layer of the **Info** pane.

22. **Right-click** over the Moon and select **Show Info** from the **Object Contextual Menu**. Make a note of the **Angular Size** of the Moon as shown in the **Other Data** layer of the **Info** pane.

Question 8. What is the angular distance between the Sun and the node of the Moon's orbit?

Question 9. What is the angular separation between the center of the Sun and the center of the Moon as seen from the center of the Earth?

Question 10. Using the angular sizes of the Sun and the Moon that you obtained from the Info pane and your answer to the previous question, what is the angular distance that separates the south limb (bottom edge) of the Moon from the north limb (top edge) of the Sun? [Hint: a simple diagram might help in this calculation.]

Perhaps you are disappointed to find that the Moon does not appear to eclipse the Sun on this date from your observing location at the center of a transparent Earth.

23. **Centre** the view on the **Sun** once more, if necessary.

24. Click the **Increase Current Elevation** button in the toolbar once, thereby moving your viewing location from the center of a transparent Earth to the surface of the Earth at the north pole.

The Moon is much closer to the Earth than the Sun and so this change of observing location of about 6400 km produces a parallax effect, which results in the Moon appearing to move relative to the more distant Sun. You can demonstrate this parallax effect by pointing your finger at arm's length at a distant object and looking at your outstretched finger with one eye closed. If you now open this eye and close the other eye, you will see that the position of your finger appears to move and is now aligned with a different position in the distance. This apparent change of the closer object as a consequence of a change of observing position is known as **parallax**. In the same way, a change in viewing location on the Earth's surface can bring the Moon into alignment with the Sun.

You can go to a location on the Moon to view the Moon's shadow on the Earth at this specific time.

25. Open the **Find** pane and click the menu button to the left of the listing for the **Moon** and select **Go There**.

26. In the **Find** pane, click the menu button for the **Earth** and select **Magnify**.

27. Select **Options > Solar System > Planets-Moons...** and click the checkbox beside the option **Outline Edge of Umbra and Penumbra** in the **Planets-Moons Options** dialog window in the section labeled **Eclipses**.

The view shows the Earth as seen from the Moon. The tan-colored line displayed on the Earth indicates the shape and location of the Moon's shadow. Positions on the Earth within this circle will experience a partial solar eclipse.

28. Select **File > Save As...** and save this view in a temporary file named **temp**.

29. Set the **Time** in the toolbar to **0:30:00 UT**.

30. Set the **Time Flow Rate** to **300×** to watch the shadow of the Moon move across the Earth.

The Moon's parallax is sufficient to cause the Sun to be partially eclipsed by the Moon as seen from the north pole of the Earth, while for observers south of this shadow, the Moon passes near, but not in front of, the Sun. You can see that Australia is well south of this shadow.

31. Select **File > Revert** to revert to the point at which you saved the view as a file named **temp**. Position the **Hand Tool** over **Australia** in the image of the Earth, **right-click**, and select **Go There** from the **Object Contextual Menu**.

32. Open the **Find** pane and select the **Centre** option from the menu for the **Sun**. You can **Zoom In** on the Sun to see the view of the Sun and the Moon as seen from a location in Australia at the same time as a partial solar eclipse is visible from the north pole. Note that these objects are well separated in the sky from this location.

B. Eclipse Seasons and the Frequency of Solar Eclipses

As the eclipses in Section A demonstrate, the Sun and Moon do not have to be exactly at a node for a solar eclipse to be visible from somewhere on the Earth. This is because the Sun and Moon are extended objects in our sky, having angular sizes of about 30'. Also, the Earth is large enough that parallax caused by changes in viewing location can bring the Moon into alignment with the Sun.

These factors set the range of positions of Moon and Sun away from the node over which an eclipse can be seen from somewhere on Earth. On average, a solar eclipse can occur whenever the Sun is within 15° of either side of one of the nodes of the Moon's orbit. The duration of this period during which the Sun is within these limits is called an **eclipse season**, as discussed earlier. An eclipse will occur while the Sun is within this range provided that the Moon reaches its new phase during this eclipse season.

33. Select **Favourites > Observing Projects > Solar Eclipses > Eclipse Season.** The view should be centered on the labeled star Rho Leonis. If this is not the case, select **File > Revert.**

The view is centered on the labeled star Rho Leonis. The descending node of the Moon's orbit is just to the left of this star. The Sun is on the ecliptic in the upper right quadrant of the view, about 15° from the node. This marks the beginning of an eclipse season, which will last until the Sun has moved beyond the node by about 15°.

34. With the **Time Flow Rate** set at **30 days, Step time forward** once and measure the angular separation between the Sun and the node.

Question 11. Approximately how long, in days, does an average eclipse season last?

Question 12. How long is a synodic month? (Hint: Check the textbook for this information.)

Question 13. Is it possible that no new Moon will occur during an eclipse season?

Question 14. Is it possible for more than one new Moon to occur during an eclipse season?

Question 15. What is the minimum number of eclipses that can occur during an eclipse season?

Question 16. What is the maximum number of eclipses that can occur during an eclipse season?

The time required between successive passages of the Sun through a specific node is called an **eclipse year**. If the Moon's orbit were fixed relative to the background stars, an eclipse year would be equivalent to a sidereal year of 365.25 days. Careful observers of the previous sequence will have noticed that during the interval of the eclipse season, the position of the descending node of the Moon's orbit shifted with respect to the background stars. This suggests that the Moon's orbit is not fixed with respect to the background stars and that an eclipse year is different in length from a sidereal or calendar year. The reason for this drift of the nodes is the slow precession of the Moon's orbit.

35. Select **Favourites > Observing Projects > Solar Eclipses > Eclipse Year.**

The view is centered and locked on the Sun as it is crossing the descending node of the Moon's orbit. The labeled star Zavijava serves as a reference star. North is at the top of the view and west is to the right.

36. Note the **Date** in Data Table1 below.
37. **Zoom Out** to a field of view 100° wide.
38. With the **Time Flow Rate** set at **1 day, Run time forward** to observe the Sun's motion along the ecliptic until it nears the ascending node of the Moon's orbit.
39. **Stop Time** and **Zoom In** on the view.
40. **Step time forward** and **Backward** to find the date at which the Sun is nearest to the ascending node. Record this **Date** in Data Table 1 below.
41. **Zoom Out** to a field of view 100° wide and **Run time forward** to observe the motion of the Sun along the ecliptic until it returns to the descending node of the Moon's orbit.
42. **Stop Time** and **Zoom In.**
43. **Step time forward** and **Backward** in intervals of **1 day** to find the date at which the Sun is closest to the descending node. Record the **Date** of this node crossing in Data Table 1 below.

Data Table 1. Dates of Node Crossings by the Sun

Node Crossing	Date
Descending Node Crossing	
Ascending Node Crossing	
Descending Node Crossing	

Question 17. How many eclipse seasons are in an eclipse year?

Question 18. How long in days is an eclipse year?

Question 19. By approximately how many days does an eclipse year differ from a sidereal year?

Question 20. Is it possible for more than two eclipse seasons to occur in the duration of a calendar year?

Question 21. What is the minimum number of solar eclipses that occur in a calendar year?

Question 22. What is the maximum number of solar eclipses that can occur in a calendar year?

C. Precession of the Nodes

In the previous section, you saw that the nodes of the Moon's orbit moved along the ecliptic. Gravitational effects from the Sun and Earth cause the Moon's orbit to precess and this causes the nodes to drift slowly over time relative to the background stars. This effect is known as the **regression of the nodes.**

44. Select **Favourites > Observing Projects > Solar Eclipses > Regression of Nodes.**

This view is from a location in space 0.005 AU over the northern hemisphere of the Earth, looking down onto the orbit of the Moon. As seen from this location, the descending node of the Moon's orbit is in the direction of the labeled star Phi Velorum. Since the view is directed toward the south, motion in a clockwise direction will be westward and motion in an anticlockwise direction will be eastward.

45. Record the **Date** for this alignment, shown in the toolbar, in Data Table 2 below.

46. Advance the **Year** and **Month** in the **Date** and watch the motion of the node along the orbit of the Moon. When this descending node (hollow wedge icon) returns to the view and approaches the star Phi Velorum, **Stop Time** and use the **Step time forward** button in 2-day steps to return the descending node to its initial position relative to the reference star. Record the **Date** shown in the toolbar in Data Table 2 below.

Data Table 2. Date of Alignment of Descending Node of Moon's Orbit with Reference Star

Date of Alignment Event #1	
Date of Alignment Event #2	

Question 23. Approximately how long, in years, does it take for the line of nodes to complete a complete cycle with respect to the star background?

Question 24. What is the angular speed of the regression of nodes in degrees per year?

D. Saros Cycle

Through observations of lunar and solar eclipses over many generations, ancient Babylonian astronomers noticed that eclipses have a natural cycle. A solar eclipse occurring in a particular constellation is followed 6585 days later by a similar eclipse near the same region of the sky. This cycle of 6585 days is known as the **Saros** cycle.

> **Question 25.** How many synodic months are there in one Saros cycle? [Hint: A synodic month is 29.53 days in length.]

> **Question 26.** How many eclipse years are there in one Saros cycle? [Hint: The precise length of an eclipse year is 346.62 days.]

The fact that the answers to the previous two questions are both very nearly integers explains the Saros cycle. In a Saros cycle, the Sun has moved through an integral number of eclipse years and has returned to the same node. Meanwhile, the Moon has completed an integral number of lunations to return to the same node in its new phase.

E. Types of Solar Eclipse

Whenever the Sun is within about 15° of a node, a solar eclipse of some kind is inevitable. The type of eclipse seen by observers on Earth will depend on when the eclipse takes place within the eclipse season. Early and late in the eclipse season, the Sun and Moon are relatively far from the node and are therefore relatively far from each other. An eclipse occurring at this time in the eclipse season will be partial. Eclipses that occur nearer the middle of the eclipse season, when the Sun and Moon are closer to the node, are more likely to have more overlap of the Moon upon the Sun. From some locations on the Earth, the centers of the Sun and Moon will appear to align perfectly and produce a **central eclipse**.

There are two types of central eclipses: annular and total. The difference between these two eclipses arises from the fact that the orbit of the Moon around the Earth and the orbit of the Earth around the Sun are elliptical in shape. The ellipticity of these orbits causes the apparent sizes of the Sun and Moon to vary, as seen from the Earth.

The radius of the Sun is nearly 696,000 kilometers, while that of the Moon is 1738 kilometers. On the other hand, the Sun is at an average distance of about 150 million kilometers from Earth, whereas the average distance from the Earth to the Moon is about 384,500 kilometers.

> **Question 27.** What is the ratio of the radii of the Sun and the Moon?

> **Question 28.** What is the ratio of the average distances of the Earth from the Sun and the Moon?

> **Question 29.** How do the two ratios compare?

As the Moon moves around the Earth in its elliptical orbit, it will be closest to the Earth at perigee and its angular size will be larger than at any other time. Conversely, the Moon's angular size as viewed from Earth will be smallest when it is at apogee.

> 47. Select **Favourites > Observing Projects > Solar Eclipses > Apparent Size of the Moon.**

The view is centered on a close up of the full Moon as seen from the center of a transparent Earth. Note the proportion of the field of view that the Moon occupies.

> 48. In order to see more clearly the change in the apparent size of the Moon as seen from Earth, you can remove the effect of the Moon's phases. Select **Options > Solar System > Planets-Moons...** from the menu and, at the top of the **Planets-Moons Options** dialog box, click the **Show Dark Side** option to uncheck it and uncheck the **Specular Reflection** option. This will result in an evenly illuminated lunar disc despite the passage of time. Click **OK.**

49. A useful aid in observing this subtle change in apparent size is a suitable **Field of View** circle superimposed on the image. Open the **FOV** pane and check the **30 Arcminutes** box to introduce this circle.

50. With the **Time Flow Rate** set to **1 hour**, **Run time forward** and watch the apparent size of the Moon change over time.

51. **Stop Time.**

52. **Right-click** over the image of the **Moon** and select **Show Info** from the **Object Contextual Menu** for the Moon. In the **Info** pane for the Moon, expand the **Other Data** layer.

53. Change the **Time Flow Rate** to **1 day**.

54. **Run time forward** and watch the value in the **Info** pane for the **Angular Size** of the Moon change over the course of at least one year, noting the minimum and maximum values.

Question 30. Approximately how many days does it take for the Moon to go through a complete cycle, from smallest apparent size through largest apparent size and back to its smallest apparent size?

Question 31. What is the approximate range of the apparent angular size of the Moon in the sky?

Just as the Moon appears alternately larger and smaller through the month due to the elliptical orbit of the Moon around the Earth, the Sun also appears larger and smaller in our sky as the Earth moves through its elliptical orbit around the Sun. When the Earth is closest to the Sun, at perihelion, the Sun will appear slightly larger in the sky than when the Earth is at aphelion, the farthest point in its orbit from the Sun.

55. Select **Favourites > Observing Projects > Solar Eclipses > Apparent Size of the Sun**. Open the **FOV** pane and click the **30 Arcminutes** box to superimpose this field of view upon the Sun's image.

56. With the **Time Flow Rate** set at **3 days**, **Run time forward** and observe the change in apparent size of the Sun in the view.

57. **Stop Time.**

58. Open the **Info** pane for the **Sun** and expand the **Position in Space** and **Other Data** layers.

59. **Run** or **Step time forward** for at least a year while watching the value for the **Distance from Observer** and the **Angular Size** of the Sun in the **Info** pane. Note the times at which these values reach their minimum and maximum values.

Question 32. During which month does the Sun appear smallest in our sky?

Question 33. During which month is the Earth at aphelion?

Question 34. In what month does the Sun appear largest in our sky?

Question 35. In what month is the Earth at perihelion?

Question 36. What is the approximate range in angular size of the Sun through the year?

Figure 3–11 in Freedman, Geller, and Kaufmann, *Universe*, 9th Ed., illustrates the shape of the Moon's shadow.

The Moon's shadow cast by the Sun is in the shape of a diverging cone (the penumbra) that becomes wider with increasing distance from the Moon. In a solar eclipse, the Earth passes into some part of the Moon's shadow cone and all or part of the Moon's circular shadow is cast on the surface of the Earth. Whenever an eclipse occurs while the Sun is within about 10° of a node, the umbra, the central converging part of the shadow, will reach the Earth and produce a central eclipse.

60. Select **Favourites > Observing Projects > Solar Eclipses > View from Cairo.**

The view is from Cairo, Egypt, and is centered on the Sun at 11:00 AM on October 3, 2005. The Moon is visible in the view, above and to the right of the Sun. The orbit of the Moon and the descending node of its orbit are also displayed.

61. Set the **Time Flow Rate** to 300× and observe the partial solar eclipse. **Stop Time** near the point of maximum eclipse and measure the angular distance from the Sun to the descending node. You can re-run this sequence after zooming in on the Sun to see the partial eclipse in more detail.

Question 37. Based on the distance of the Sun from the node, is this likely to be a central eclipse from some location on Earth?

In the next sequence, you will observe this same event from a position in space about 0.001 AU over Cairo, looking down on the Earth.

62. Select **Favourites > Observing Projects > Solar Eclipses > Cairo from Space.**

The view is looking down on Earth from a position directly over Cairo, Egypt. In the upper left quadrant of the Earth, the outline of the Moon's penumbral shadow is illustrated.

63. Select a **Time Flow Rate** of 300× and observe the Moon's shadow track across Africa.

As you learned in the previous section, the Sun needs be only within about 15° of one of the nodes of the Moon's orbit for a solar eclipse to occur. Whenever the Sun is within about 10° of a node at the time of new Moon, the alignment of the Moon and Sun as seen from Earth is close enough that the central portion of the moon's circular shadow touches some point on the Earth.

The central part of the shadow cone, the umbra, is the darkest. From anywhere within the umbra, the Moon would be seen to blot out the sun completely. Surrounding the umbra is the penumbra, where light from some portion of the Sun is visible, producing a solar eclipse that is only partial (i.e., the Moon covers only part of the Sun).

The average length of the Moon's umbral shadow is somewhat shorter than the average distance from the Moon to the Earth's surface. When the Moon is far enough away that the umbra does not reach the Earth, the Moon appears smaller in the sky than the Sun, and the entire shadow on the Earth is penumbral; (i.e., a portion of the Sun is visible from all points in the Moon's shadow). At maximum eclipse, a thin, bright ring of the Sun's disk surrounds the occulting Moon. This type of central eclipse is called an **annular eclipse**.

64. Select **File > Revert.** Then change the **Time** in the toolbar to **1:02:00** PM. You will notice that the Moon's shadow as represented by the tan-colored oval contains a darker region near its center.

65. Position the mouse cursor over the central part of this darker region of the Moon's shadow and open the **Object Contextual Menu** over this location on the Earth. Select **Go There** from the menu to go to the surface of the Earth at the Latitude and Longitude specified in the menu option.

66. Open the **Find** pane, click the menu button to the left of the entry for the **Sun** and select **Magnify**. Adjust the **Time** in the toolbar to the time of maximum eclipse.

67. Adjust your viewing location to align the center of the Sun and the Moon more precisely by selecting **Options > Viewing Location...** and then clicking the **Latitude/Longitude** tab in the dialog window. Set the **Latitude** to **24 N** and the **Longitude** to **19.5 E**. Then click the **Set Location** button.

68. Open the **Find** pane, click the menu button to the left of the entry for the **Sun** and select **Magnify**. Adjust the **Time** in the toolbar to the time of maximum eclipse. The view shows the eclipse as it is seen from within this darker region of the Moon's shadow. Again, you can **Zoom In** on the Sun to see this eclipse more closely.

69. Click the menu button for the **Sun** in the **Find** pane and select **Show Info.** Note the **Angular Size** of the Sun as seen from Earth on this date under the **Other Data** layer of the **Info** pane.

70. Open the **Find** pane and click the menu button for the **Moon** and select **Show Info.** Note the **Angular Size** of the Moon as seen from Earth on this date under the **Other Data** layer of the **Info** pane.

You are now viewing the same eclipse that you observed as a partial solar eclipse from Cairo, Egypt, in a previous sequence but from a slightly different location.

> **Question 38.** What type of eclipse occurs, as seen from this location, within the darkest part of the Moon's shadow?

F. Total Solar Eclipses

During some eclipses, the apparent sizes of the Moon and of the Sun and their alignment are such that the Moon's umbra reaches all the way to the Earth. As the Moon orbits eastward around the Earth, the umbra traces a path across the Earth's surface from west to east. For observers located along the centerline of the shadow, the Moon completely blots out the bright disk of the Sun at the midpoint of the eclipse. These observers witness a **total solar eclipse,** one of the most awesome sights of nature. Many people travel around the world to be on the centerline at this time just to experience this rare event.

There are four distinct contact points in a total eclipse. **First contact** occurs when the Moon first encroaches on the Sun, as the edge of the Moon's penumbral shadow just begins to cross the viewing location. After first contact, the Moon gradually covers more and more of the Sun's disk during the partial phase of the eclipse. As seen from space, the viewing location on the Earth becomes more deeply engulfed in the penumbral shadow of the Moon. Only when the Moon has covered about 90% of the Sun's disk will observers on the ground notice a significant diminution in the brightness of the daylight. The sky becomes a very dark twilight blue. When only the smallest sliver of the sun's disk remains uncovered, the topography of the Moon becomes evident as irregularities in the slender remaining crescent part of the Sun when viewed through a dark filter. As the Moon continues toward a complete occultation of the Sun, a few last rays of sunlight may pass through lunar valleys and strike the Earth. The result, as seen from the Earth, is a string of bright beads of sunlight along the advancing limb of the Moon. These are called **Baily's Beads.** When only one such bead remains, the shadow has deepened sufficiently that the entire disk of the Moon appears encircled by the silvery glow of the solar corona, while the bright solitary bead of photosphere sparkles like a diamond. The effect is called, aptly, the **diamond ring effect.**

At **second contact,** the Moon completely covers the Sun's disk and the marvel of totality unfolds. Only at this time is it safe to look directly at the Sun without eye protection, and one is encouraged to do so. With the bright photosphere of the Sun completely blocked out by the Moon, the relatively fainter atmospheric layers of the sun, the chromosphere and corona, become visible. As seen from space, locations on the Earth that lie in the path of totality are within the Moon's umbral shadow.

At **third contact,** totality ends and it is once again important to use proper eye protection while observing the final partial phases of the eclipse. The diamond ring effect occurs when the first bead of photosphere is uncovered, followed by Baily's Beads and then the final partial phases of the eclipse. At **fourth contact,** the Moon no longer occults any part of the Sun.

In the next sequences, you will observe the total solar eclipse that occurred on July 11, 1991. This eclipse was notable for its long duration as well as for the fact that the centerline of totality passed some of the most modern astronomical observatories in the world atop Mauna Kea in Hawaii.

71. Select **Favourites > Observing Projects > Solar Eclipses > View from Mauna Kea.**

The view is centered on the Sun as seen from Mauna Kea in Hawaii on July 11, 1991, just after sunrise.

72. Set the **Time Flow Rate** to 300 × and watch this eclipse.

73. Select **File > Revert.**

74. **Zoom In** to a field of view about 3° wide. Manipulate the time flow controls and obtain times for each contact point of this solar eclipse.

This sequence, particularly around totality, provides an excellent simulation of a total solar eclipse, including the appearance of bright stars and then fainter stars as the sky darkens, followed by the slow unveiling of the faint corona that can be seen to surround the Sun during totality. While a simulation cannot come close to showing the beauty and majesty of a real eclipse, it nevertheless captures the essence of the event as it unfolds for an observer within the path of totality.

Question 39. From this location, when do the following events occur?

        a) First contact

        b) Second contact

        c) Third contact

        d) Fourth contact

Question 40. What is the duration of totality at this location?

The duration of totality will be longer, the closer the viewing location is to the center line of the eclipse track. The next sequence displays the same eclipse from a location about 105 kilometers southeast of Mauna Kea Observatory.

75. Select **Favourites > Observing Projects > Solar Eclipses > View From 19N 155W.**

76. **Zoom In** to a field of view about 3° wide.

77. Manipulate the time controls and obtain the times of each contact of the eclipse event as seen from this location.

Question 41. What is the duration of totality at this location?

Question 42. Which of these two locations, Mauna Kea or the site SE of the observatory, is nearer to the center line of the shadow track?

G. Maximizing Totality

The stunning beauty of totality prompts eclipse chasers to try to maximize the time they spend in the umbral shadow of the Moon. Choosing a location as close to the centerline of the eclipse track is important, but other factors also play a role. The closer to a node that an eclipse occurs, the more closely the Sun and Moon will align, increasing the duration of the eclipse. The relative apparent angular sizes of the Sun and the Moon as seen from Earth can also affect the duration of a total solar eclipse. The nearer the Moon is to perigee, the larger it will appear. When the Earth is at aphelion, the Sun will appear relatively smaller than when the Earth is at perihelion.

78. Select **File > Revert**.

79. **Run time forward** to the time of maximum eclipse at the midpoint between second and third contact.

80. Open the **Find** pane and click the box to the right of the label for the **Moon** to display its orbit. In addition to the wedge icon indicating the position of the node, the display of the Moon's orbit also shows the point at which the Moon is at perigee, indicated by a short line perpendicular to the orbit.

Question 43. List the factors that contribute to the relatively long duration of the July 11, 1991, total solar eclipse.

In addition to choosing a location on the Earth as close to the centerline of the eclipse as possible to extend the duration of totality, choosing a location where the eclipse occurs near midday and near the Earth's equator will also significantly increase the duration of totality. At midday, the Sun transits the observer's meridian, and since the Sun is being eclipsed by the new Moon, the Moon is also on the meridian at midday. The Moon appears slightly larger when it is on the meridian than when it is seen at the time of moonrise or moonset because the rotation of the Earth on its axis has carried the observing location closer to the Moon by a distance of almost the radius of the Earth, if the location is also near the equator.

81. Select **Favourites > Observing Projects > Solar Eclipses > View From Mexico**. Examine the view and the values displayed in the toolbar.

The date is July 11, 1991, and this is the same eclipse that you observed from Hawaii in the last two sequences. Note that the local time is near midday at this location in Mexico.

82. Position the **Hand Tool** over the image of the Moon again to open the **Object Contextual Menu** and select **Show Into**. In the **Info** pane, expand the layers named **The Moon Info** and **Other Data**.

83. Position the **Hand Tool** over the image of the Moon again and in the **Info** pane, expand the layers named **The Moon Info** and **Other Data**.

84. In the **Other Data** layer, note the apparent angular size of the Moon in the sky at this time near midday.

85. Click the button to the right of the **Rises** label in **The Moon Info** layer of the **Info** pane to change the time to that of moonrise. Notice the apparent angular size of the Moon at this time. Click the button to the right of the **Sets** label in the **Info** pane layer and notice the apparent angular size of the Moon at the time of its setting.

86. Select **File > Revert**.

87. Use the **Time Flow** controls in the toolbar to determine the times of each contact point of the eclipse as seen from this location.

Question 44. What is the duration of totality at this location?

Choosing to observe a solar eclipse from the location on the Earth over which the center of the Moon's shadow passes at midday extends the duration of totality for another and more significant reason than the slight increase in angular size of the Moon when it is observed near to the meridian. The rotation of the Earth will carry this observing location more rapidly along the eclipse track in the same direction as the motion of the Moon's shadow and will thus increase significantly the duration of totality. During a total eclipse observed nearer to sunrise or sunset, the Earth's rotation is carrying the observing location nearly directly toward or away from the Moon, while the orbital motion of the Moon around the Earth carries the umbra eastward across the surface of the Earth. Someone observing the same eclipse from a location at which the umbra passes near midday is carried in a more direct easterly direction by the rotation of the Earth and so manages to keep up with the shadow for a longer duration.

88. Select **File > Revert**.

89. Position the **Hand Tool** over the center of the image of the Moon, open the **Object Contextual Menu** and select **Go There** to move to a viewing location on the Moon.

90. Open the **Find** pane and click the menu button to the left of the entry in the list for the **Earth** and select **Magnify**.

This view shows the Earth centered in the view as seen from the Moon. Toward the left (west) limb of the Earth, a relatively large semicircle indicates the intersection of the Moon's penumbral shadow with the surface of the Earth. Within this region, a small circular outline near the west limb of the Earth shows the size and position of the umbra of the Moon's shadow.

91. Set the **Time Flow Rate** to 300× and observe the track of the Moon's shadow, particularly that of the umbra, in the July 11, 1991, eclipse.

92. Select **Edit/Undo Time Step** from the menu to return to the time at which the umbra of the Moon's shadow is near the west limb of the Earth as seen from the Moon. Recall that for observers within the umbral shadow at this time, the Sun has just risen.

93. Select **Options > Solar System > Planets Moons...** from the menu. In the **Planets-Moons Options** dialog box, click the **Surface Guides** option to check it and then click the **Grid** option.

94. **Run time forward** at 300× to replay the animation, watching the relative speed of the Moon's umbra across the surface of the Earth and paying particular attention to its speed for observers who would be seeing the eclipse near sunrise and sunset, when the umbra is near a limb of the Earth, and for observers who would see the eclipse near midday, when the umbra tracks across the center of the Earth.

Question 45. In which direction across the Earth is the Moon's shadow moving?

Question 46. In which direction is the Earth rotating?

H. The Shadow Track

In this section, you will use *Starry Night Enthusiast*™ to examine the last solar eclipse over Europe in the twentieth century, which occurred on August 11, 1999. The eclipse started over the Atlantic Ocean. The shadow then moved quickly over the southern tip of England, over northern France, Germany, Austria, Romania, and out onto the Black Sea before covering a strip of Turkey, continuing over Asia and ending at sunset in the Bay of Bengal on the Indian Ocean. The greatest width of the path of totality was only 112 km, and this occurred over the Black Sea. The maximum duration of totality was only 2 minutes, 23 seconds, in Romania. You can first watch the shadow path of this eclipse, as it would be seen from the Moon.

95. Select **Favourites > Observing Projects > Solar Eclipses > Aug 11 Eclipse Track.**

The observing location is near Bruce Crater on the Moon. The date is August 11, 1999.

96. If you desire, activate the surface guides on the Earth by selecting **Options > Solar System > Planets-Moons > Surface Guides.**

97. With the **Time Flow Rate** set at **20 seconds, Run time forward** and watch the Moon's shadow move across the Earth.

98. Select **File > Revert** from the menu and then **Run time forward** until the small circle outlining the Moon's umbra crosses the upper-center of the image of the Earth and **Stop Time.** Zoom In and position the cursor over the center of the circle outlining the Moon's umbral shadow. Open the **Object Contextual Menu** over this spot and select **Go There.**

99. Open the **Find** pane. Click the menu button to the left of the entry for the **Sun** and select **Magnify.**

100. In the **Find** pane, click the check box to the right side of the entry for the Moon to display the Moon's orbit in the view. **Zoom Out** until the node at which this eclipse occurs is visible in the view. Measure the angular distance between the Sun or Moon and the node.

101. Click the menu button next to the entry for the **Moon** in the **Find** pane and select **Show Info.**

Question 47. If you were to observe this eclipse from a location on Earth where the umbra passes over the location near midday, would you expect totality to be longer or shorter than the view of the July 11, 1991, eclipse from Mexico? Explain your answer using the data available from the Info pane and the toolbar.

Note the small size of the umbral shadow compared to that for the July 11, 1991, eclipse—a fact reflected in the difference in the period of totality in the August 11, 1999, eclipse.

You can estimate the speed of the eclipse shadow by timing the simulated eclipse at two different sites, one just south of England, the other in the city of Munich, Germany, some 530 miles, or 850 km, apart.

102. Select **Favourites > Observing Projects > Solar Eclipses > Aug 11 from English Channel.**

103. **Run time forward** and record the times of second and third contact.

104. Select **File > Revert** and replay the sequence but **Stop Time** at second contact and note the **Time** of this event in the toolbar.

105. Select **Favourites > Observing Projects > Solar Eclipses > View from Munich.**

106. Keeping in mind the fact that Munich is one time zone east of the English Channel location, adjust the **Time** in the toolbar so that it is simultaneous with second contact at the English Channel location.

107. **Run time forward** and determine the times of second and third contact as seen from Munich.

Question 48. What is the time of mid-eclipse (calculated from the measurements of times of second and third contacts) for both sites?

Question 49. Using the two times of central eclipse at these two locations, allowing for the time zone change (subtract 1 hour from your difference), and the distance between the English Channel site and the Munich site of some 530 miles (850 km), what is the speed of the shadow of the Moon as it moves across this area of northern Europe?

I. Conclusions

In this observing project, you have observed several solar eclipses from a variety of locations on the Earth. You have observed the shadow of the Moon pass across the Earth as seen from the Moon and from space. You have used your observations to determine the requirements for solar eclipses to occur. Finally, you have explored the factors that influence the duration of an eclipse and measured the speed at which the Moon's shadow moves across the Earth.

Phases of Venus   12

Galileo's discovery of the variable phases of Venus provided a turning point in our understanding of the universe. Prior to Galileo's observations, there were two competing models of the observable universe. The accepted model was **geocentric,** with the Earth at its center. Several philosophers from the Greek era onward had suggested **heliocentric** models, where the Sun was assumed to be at the center of the universe. Prior to Galileo, there were no direct observations to allow anyone to differentiate between these two models.

> The geocentric model of the universe is discussed in Section 4–1 of Freedman, Geller, and Kaufmann, *Universe*, 9th Ed.

At a time when the geocentric model was accepted as the correct view of the world, Galileo's observations of Venus with his newly invented telescope provided decisive support for the heliocentric theory.

The geocentric model had the Moon, the Sun, the planets, and all of the stars moving around a fixed Earth. Part of the reluctance of people to accept any model other than this one arose from the fact that they could not sense the motion of the Earth, either its rotation or its motion through space. In this model, the Moon was the closest object to the Earth, and the others in order of increasing distance were Mercury, Venus, the Sun, Mars, Jupiter, Saturn, and finally the "sphere of the stars."

In the simplest geocentric picture, each object moved in a constant direction along a circular path at a constant distance from the Earth. This model did not duplicate the observations, however, particularly the observation of **retrograde motion** of planets. Here, the normal eastward motion of planets in our sky is interrupted occasionally by westward motion relative to the background stars. Various complications were added to this early geocentric model in order to reproduce the observed motions more accurately, most notably by Ptolemy around 140 AD. The most important of these additions was the assumption that each planet moved around a smaller circle called the **epicycle,** the center of which moved along a larger circle around the Earth called the **deferent.** This refined model worked very well in predicting planetary motions over short times but became increasingly inaccurate in the centuries following its development.

In the case of Venus, this epicyclic motion brought the planet alternately closer to the Earth and farther away again, but it did not cause Venus to cross the deferent of the Sun. In this model, therefore, Venus was always closer to the Earth than was the Sun, even at its farthest distance from the Earth.

> Section 4–2 of Freedman, Geller, and Kaufmann, *Universe*, 9th Ed., discusses the heliocentric model.

As a consequence, since Venus never strays far from the Sun in the sky, we should always be looking mostly at the dark side of Venus, and we should never see more than a small part of the sunlit side along one edge of the planet. The apparent size of Venus should change as the planet moves around its epicycle due to its changing distance from the Earth, but we should always see Venus in a crescent phase, or in a "new" phase (entirely dark), when it passes close to the Sun in the sky.

In the heliocentric theory, Venus orbits the Sun. Its orbit is smaller than that of the Earth so we never see Venus stray far from the Sun in the sky. However, it can be anywhere in this orbit as seen from the Earth. If Venus is between the Earth and the Sun, we will see a crescent phase. At this time, it will be closer to us and appear larger. When it is on the opposite side of the Sun from the Earth, we will see it fully sunlit, in its full phase. At this time, it will be more distant and therefore appear smaller. If the heliocentric theory is correct, we should then see a full range of phases from crescent to full and back again as Venus orbits the Sun. Furthermore, the phases should be strongly correlated with the apparent size of the planet—largest at crescent phase and smallest at full phase. It was the observation of these correlated changes, along with several other crucial measurements that convinced Galileo that the planets orbited the Sun.

> Figure 4–14 of Freedman, Geller, and Kaufmann, *Universe*, 9th Ed., shows how Venus' phases are related to its position in its orbit, as seen from the Earth.

In this project, you can investigate Venus's motion in the sky, and see how this motion correlates with Venus' phases and apparent size.

A. Motion of Venus in Relation to the Sun

We will look toward the Sun from an elevation of 54,000 km above the Earth's north pole in this simulation. This elevation allows us to watch the behavior of the Sun, stars, and planets over the course of time without the Earth or the Moon obstructing our view. We are sufficiently close to the Earth at this altitude that the view of Venus' motion will be equivalent to that seen from the Earth's surface.

> 1. Launch *Starry Night Enthusiast*™.
> 2. Select **Favourites > Observing Projects > Phases of Venus > Venus' Motion.**

The gaze is centered on the Sun, and you can see several planets on the screen.

> 3. Select **Labels > Planets-Moons.**

Venus and Mercury are just to the right of the Sun, while Saturn is near the left edge of the screen, just below the cluster of stars known as the Pleiades. Jupiter may be just off the left edge of the view, and Uranus may be near or just off the right edge.

> 4. Select **Labels > Planets-Moons** to turn the labels off for all objects except for the Sun and Venus.
> 5. **Run time forward.**

With the gaze locked on the Sun, you can tell that the Sun moves against the background stars by the way these stars move continuously behind the fixed Sun. This apparent motion of the stars is an illusion caused by the fact that we are viewing the Sun from a moving Earth. You get a similar illusion if you stand on the edge of a merry-go-round and fix your eyes on an object at the center—it stays fixed in your vision, while the rest of the world seems to spin past it.

Venus can be seen to move with respect to the Sun, reversing its direction regularly to remain relatively close to the Sun during this motion. Observe particularly the relative speed of this motion as Venus moves from left to right, compared to its motion when moving from right to left.

Question 1. Approximately how far does Venus move away from the Sun, in degrees, before turning around and coming back toward the Sun again? For reference, the total angular width of the screen from one side to the other is 100°.

Question 2. Is the apparent speed of Venus on the screen the same when it is traveling eastward (from right to left) past the Sun as when it is traveling westward (from left to right)? (*Note:* Make sure that time is running forward when you determine the answer to this question.)

B. The Ecliptic

You may notice that Venus' motion around the Sun is slightly reminiscent of a figure eight, with a looping path on each side of the Sun. This motion seems to contradict the idea that Venus, as with all planets, orbits the Sun in a single plane.

In fact, the apparent looping motion of Venus is an illusion caused by the tilt of the Earth's equator by 23 $\frac{1}{2}$° relative to its orbital plane. The edge-on view of the orbital plane and the apparent path of the Sun around our sky over a year is known as the **ecliptic;** thus, the ecliptic is tilted by 23 $\frac{1}{2}$° relative to the celestial equator. Our view from above the Earth's north pole is oriented with the top and bottom of our view parallel to the celestial equator, and the left and right sides parallel to the Earth's rotation axis. As our view follows the Sun around the ecliptic each year, the ecliptic appears tilted from lower right to upper left when we look toward the Vernal Equinox in March, and from lower left to upper right six months later, when we look toward the Autumnal Equinox in September. Thus, although the celestial equator and the ecliptic are fixed planes, our changing direction of view causes the ecliptic to appear to sweep through + and −23 $\frac{1}{2}$° over the course of a full year. Venus' orbit is almost in the plane of the Earth's orbit, so this orbit will also tilt back and forth along with the ecliptic plane.

> 6. Select **View > The Ecliptic.**

With time progressing forward, you can see that the Sun follows the ecliptic around the sky, and that Venus stays close to the ecliptic (that is, close to the plane of the Earth's orbit) at all times. The small departures of Venus from the ecliptic are caused by the fact that Venus' orbit is inclined by a small angle of about 3° to the Earth's orbit.

> 7. Activate the **Celestial Grid** in the **View** menu.

The celestial grid (also known as the **equatorial coordinate system**) is a coordinate system on the sky, based on the Earth's equator and its lines of latitude and longitude. The horizontal lines denote **declination** lines, which are equivalent to lines of latitude on the Earth and are parallel to the equator. We can use these lines as a reference to monitor the change in the tilt of the ecliptic. In this wide-angle view of the sky, the ecliptic appears to be curved. Nevertheless, you can still follow the changing angle of the ecliptic through +/− 23 $\frac{1}{2}$° with respect to the declination lines as time progresses through the year.

> 8. Open the **Find** pane and click the checkbox to the right of the listing for Venus to display its orbit. Then close the **Find** pane.
> 9. **Run time forward** and **backward** to see Venus following its orbit around the Sun.
> 10. **Stop Time** and change the Date in the toolbar to June 7, 2001.

Venus' orbit is edge-on on this date and intersects the ecliptic at the position of the Sun. Venus can be seen at the extreme right-hand side of its orbit at this time, a position known as **greatest western elongation.** You can see from this fortunate alignment that the angle between Venus' orbital plane and the ecliptic is a few degrees.

C. The Phases of Venus

> 11. Select **Favourites > Observing Projects > Phases of Venus > Venus' Phases.**

The view on the screen is similar to that in part A, except that the view is now locked on Venus. With Venus locked at the center of the screen, it becomes easy to zoom in to see the phase of Venus and then zoom back out again to see Venus' position and motion relative to the Sun.

> 12. **Run time forward.**

The view on the screen looks different from that in part A because the view is locked on Venus. Consequently, the Sun now appears to orbit around Venus. This is an illusion caused by fixing our gaze on a moving planet. The relative orientation of Venus and the Sun remains the same as in part A, with Venus sometimes to the left and sometimes to the right of the Sun.

> 13. Select **File > Revert.**
> 14. Display Venus' orbit by right-clicking over the planet and selecting **Orbit** in the object contextual menu. This will help you to visualize the motion of Venus around the Sun.

In this part of the project, you will observe Venus at various points in its motion relative to the Sun. At each point you will:

Record the date

Measure the angular separation between the Sun and Venus

Note the distance to Venus from the observing location near the Earth

Observe and record the phase of Venus

Determine the angular size of Venus in the sky

Note that, on April 5, 2001, Venus is close to the Sun in the sky.

> 15. Use the **Hand Tool** to measure the angular separation between the Sun and Venus and note the **Date** and the Venus–Sun separation in Data Table 1 below. **TIP:** It is easier to start at the Sun when using the **Hand Tool** to measure this separation. If you accidentally scroll the view while making this measurement and lose the lock on Venus, select **Edit > Undo Scroll** from the menu.
> 16. Open the **Info** tab for Venus. Under the **Position in Space** layer, find the value given for **Distance from observer** and record this into the Data Table below, in the **Earth–Venus Distance** column.
> 17. **Zoom in** on Venus to a field of view about **6'** wide.

You should now see a close-up of Venus, showing its phase. Venus should appear as a very thin crescent. You see the view that is observed through a telescope from Earth, the top of the thick atmosphere surrounding Venus. (Note: You can remove this atmosphere at any time to reveal the Venus surface detected by spacecraft measurements by selecting **Options > Solar System > Planets-Moons** and clicking the **Show Atmosphere** box under the **Surface** heading.)

> 18. Record the phase of Venus for this date in the Data Table.
> 19. Under the **Other Data** layer in the **Info** pane, find the **Angular size** of Venus in the sky and record this value in the Data Table.
> 20. **Zoom out** again to the widest field of view.

Data Table 1. Phase and Angular Size of Venus

Date	Venus–Sun Separation (° ' ")	Earth–Venus Distance (AU)	Phase of Venus	Angular Size of Venus (")

In the steps below, you can watch the phase of Venus change as it moves around the Sun.

21. Ensure that the view is still centered (i.e., locked) on Venus and **Run time forward** to **May 8, 2001.** Repeat the measurements and observations described in Steps 15–20 for this date and record your results in the Data Table above.

Question 3. How has the apparent angular separation between Venus and the Sun changed between April 5, 2001, and May 8, 2001?

Question 4. How has the phase of Venus changed between these dates?

Question 5. Has the apparent size of Venus increased, decreased, or stayed the same from April 5 to May 8, 2001?

22. **Run time forward** to **June 7, 2001.** Repeat the measurements and observations described in Steps 15–20 for this date and record your results in the Data Table above.

Venus is west of the Sun (toward the right on the screen) and should now be at about its greatest angular distance from the Sun. This places Venus near **greatest western elongation.**

Question 6. What phase does Venus show when it is near greatest western elongation?

Question 7. Has the apparent size of Venus increased, decreased, or stayed the same from May 8, 2001, to June 7, 2001?

23. **Run time forward** to **December 21, 2001.** Repeat the measurements and observations described in Steps 15–20 for this date and record your results in the Data Table above.

By December 21, 2001, Venus is close to the Sun again in our sky and moving rapidly toward **conjunction,** the time at which the angular distance between Venus and the Sun is smallest. At this time, Venus is beyond the Sun and at its farthest distance from Earth and hence will appear to be at its smallest at this point.

Question 8. How has the phase of Venus changed between greatest western elongation and conjunction?

Question 9. Has the apparent size of Venus increased, decreased, or stayed the same from greatest western elongation to conjunction?

24. **Run time forward** to **August 16, 2002.** Repeat the measurements and observations described in Steps 15–20 for this date and record your results in the Data Table above.

By August 16, 2002, Venus is again near its farthest position from the Sun in the sky, this time to the east of the Sun (toward the left side of the screen); i.e., Venus is now at **greatest eastern elongation.**

> **Question 10.** How has the phase of Venus changed between conjunction and greatest eastern elongation?
>
> **Question 11.** Has the apparent size of Venus increased, decreased, or stayed the same from conjunction to greatest eastern elongation?

25. **Run time forward** to **October 21, 2002.** Repeat the measurements and observations described in Steps 15–20 for this date and record your results in the Data Table above.

By October 21, 2002, Venus is again approaching conjunction with the Sun, this time from the east.

> **Question 12.** How has the phase of Venus changed between greatest eastern elongation and conjunction?
>
> **Question 13.** Has the apparent size of Venus increased, decreased, or stayed the same from greatest eastern elongation to conjunction?

The data in the Data Table show you the dependence of angular radius on the Earth–Venus distance and demonstrate the large change in angular size of Venus during this relative motion of Earth and Venus.

The following steps allow you to see the progression in Venus' phase and apparent size in the sky even more clearly than the steps above.

26. Select **File > Revert.**
27. **Zoom in** to a field of view that is 3' wide.
28. **Run time forward.** Allow time to progress until at least **December 20, 2002** to show the dramatic changes in the size and phase of Venus during this full cycle.

> **Question 14.** What relationship is there between the "fullness" of the phase of Venus (e.g., thin crescent, thick crescent, half full, gibbous, full) and the apparent size of Venus?
>
> **Question 15.** Do your observations support the geocentric or the heliocentric theory of the universe?
>
> **Question 16.** Which specific observations support your answer to the previous question?

D. Conclusions

With the observations and measurements in this project, you have demonstrated the large variation in angular radius of Venus that was seen by Galileo almost 400 years ago with his primitive but very effective telescope. It was on these measurements that he based his conclusion that Copernicus was correct in believing that the planets orbited the Sun rather than the Earth.

Planetary Motion: 13
The Retrograde Motion
of Mars

If we could observe an orbiting planet from a stationary position near the Sun, we would see this planet moving relatively uniformly eastward (i.e., from west to east) against the background sky. There would be small variations in speed caused by the orbit being an ellipse and not a perfect circle, but these variations are relatively small. This eastward motion is called **direct** motion.

In reality, however, we watch a moving planet from a moving Earth, and this leads to a complicated path of the planet across our sky. Consider Mars, for example. Mars is a superior planet with an orbital radius greater than that of the Earth, and thus it moves more slowly in its orbit than does the Earth. For most of the year we are either on the far side of our orbit from Mars or we are traveling more-or-less toward or away from Mars, and in these cases we see Mars moving "forward" in its orbit, i.e., eastward, or direct motion. To visualize this, imagine yourself in a fast car driving on the inside of a circular track while a slower car drives on the outside of the track. From most of the track, you will see the slower car moving in the same direction as you against the background.

However, as you overtake the other car, it will appear to drop back relative to you against the background. The same will be true for Mars as seen from a moving Earth. For part of the year we are on the same side of the Sun as Mars, and overtaking it. Because we are pulling ahead, Mars appears to drop back in a westward direction as seen from the Earth. This temporary westward motion for planets is known as **retrograde** motion. We see retrograde motion for all other superior planets for the same reason.

> Direct and retrograde motion are discussed in Section 4–1 of Freedman, Geller, and Kaufmann, *Universe*, 9th Ed.

You can demonstrate this motion easily with *Starry Night Enthusiast*™ by watching the relative motion of Mars night by night against the background stars. You will find that this reversal of apparent motion takes place when the planet is on the opposite side of the Earth from the Sun, around the position known as **opposition**. This is expected from the argument above because it is at opposition that the Earth is overtaking the more distant and hence slower planet.

> Opposition, conjunction, and other planetary configurations are discussed in Section 4–2 of Freedman, Geller, and Kaufmann, *Universe*, 9th Ed.

The time taken for the Earth-Mars system to move from one opposition to the next is known as the **synodic period** of Mars. This differs from the more fundamental revolutionary period of Mars with respect to the stars, its **sidereal period**. The latter cannot be measured directly from Earth, but its value can be determined from a simple calculation after measurement of the synodic period, as is shown later in this project.

A. Mars' Motion in the Sky

The particular period chosen for this simulation includes the close approach of Mars to the Earth in late 2007. This approach will not be as close as the previous two oppositions in 2003 and 2005. In fact, the earlier of these oppositions brought Mars closer to Earth than it had been in several thousand years as a consequence of the elliptical orbits of these two planets. (After exploring this project, you might like to demonstrate that this is in fact the case, particularly with the view of the relative orbits of Earth and Mars from space, in Section C.)

You will begin by observing the motion of Mars as seen from the Earth as time advances in sidereal days, with daylight on, as would be seen if one were observing Mars every night for several months, starting on September 23, 2007. You can then explore the change in angular size of Mars as it moves closer to Earth during this close approach. Since time is advancing in sidereal days, these observations start in the early morning hours just before sunrise and slowly move to late evening because one sidereal day is several minutes shorter than 1 solar day.

1. Launch *Starry Night Enthusiast™*.

2. Select **Favourites > Observing Projects > Planetary Motion > September 2007**.

3. Open the **Preferences** dialog, click **Show Info in Upper Left Corner of the Screen**. Select **Cursor Tracking (HUD)** in the drop-down list; in the **Show** list, check the options for **Altitude, Azimuth, Name, Object type**, and **RA (J2000)**.

In the early morning hours of September 23, 2007, in Montreal, Canada, it is already twilight and the sky appears dark blue. Mars is faintly visible, and labeled, near the center of the screen, along with a reference star, *Kappa Aurigae*. Relative to the horizon, this star is almost due south. The following steps allow you to verify this.

4. Use the **Hand Tool** to drag the screen up to show the compass points along the horizon, and check the position of Mars relative to the south point. Move the cursor over Mars to activate the **HUD** on the screen.

Question 1. What is the position of Mars, represented by azimuth and altitude, at this time?

5. Use **File > Revert** to return to the original gaze direction.

You can **Run time forward** to demonstrate the complex motions of Mars in our sky, using sidereal-day steps to maintain a fixed background of stars against which Mars' progress can be followed. For this initial investigation, we will leave daylight on to show the real appearance of the sky through the next few months, until Mars moves into evening twilight. Later, we will turn off the daylight to follow Mars' progression over a longer period.

6. **Run time forward** to **March 23, 2008** AD. (If the motion of Mars is too rapid, adjust the speed of the display by opening the **Preferences** dialog, selecting **Responsiveness** in the drop-down box and changing the **Requested Frame Rate**.)

7. Select **Edit > Undo Time Flow** and then click **Run time forward** to repeat the animation.

If you watch the time display, you will notice that the time of night becomes earlier as the days pass. This is because 1 sidereal day is about 4 minutes short of 1 solar day, as mentioned above. On September 23, 2007, Mars is just visible in the morning twilight at about 6 AM. As the days pass, you see morning twilight deepen into night and then eventually brighten into evening twilight. By March 23, 2008, the sidereal-day steps have moved the observing time to about 6 PM and Mars is beginning to fade into evening twilight.

Mars' motion is initially **direct** motion to the east, but it soon reaches a stationary point where it stops and reverses its direction. It then moves westward for a period of **retrograde** motion. Eventually, as evening twilight begins to brighten the sky, Mars reaches a second stationary point where its retrograde motion ceases and **direct** or easterly motion resumes. An observer intending to watch this period of retrograde motion would therefore have to begin to observe in the early morning hours in September 2007 and end this series of observations in the early evening hours of March 2008.

8. Use **File > Revert** to reset the view.

9. Select **View > Celestial Grid** to display the celestial coordinate system.

10. **Run time forward** to observe Mars' motion again but near each stationary point **Stop Time** and use **Single Step Forward** and **Backward** at each stationary point to determine the date and note its approximate right ascension, **RA (J2000)** in the **HUD**, to answer the following questions. You can use the **Zoom** function and the single sidereal-day steps to increase the precision of these measurements.

Question 2. What are the dates of the stationary points of Mars' motion during this retrograde period?

Question 3. How long does the period of retrograde motion last?

Question 4. What are the right ascensions of these stationary points?

11. In order to answer the following questions, use **Options > Solar System > Planets-Moons…** and in the dialog window that opens, look at the section labeled **Other** and activate the **Enlarge Moon Size at Large FOVs** option. Click **OK**.

12. Use the time controls to move the **Date** in sidereal day steps to **December 23, 2007**, near the midpoint between the two stationary points, on a day when the Moon is near to Mars in the sky. **Zoom In** to examine the Moon and note particularly its phase. Consider what this phase indicates about the position of the Sun relative to the Earth and Mars at this time and date.

13. Select **Options > Solar System > Planets-Moons…** and in the dialog, deactivate the **Enlarge Moon Size at Large FOVs** option.

Question 5. Where would the Sun be at this time, relative to the Earth and Mars?

Question 6. What would be the configuration of Mars at this time (e.g., conjunction, maximum elongation, opposition, etc.)?

In this simulation, you advanced time in sidereal-day steps to maintain the constant background of stars. This led to a uniform shift in the time of observation every night, such that twilight and eventually daylight rendered Mars invisible at either end of the observing period. This limited your view of the motion of Mars to a short period around its retrograde phase. In *Starry Night Enthusiast*™, you can turn off the daylight and extend the run time to observe the apparent motion of Mars over a longer period to see more of the direct motion.

14. Select **Favourites > Observing Projects > Planetary Motion > July 2007.**

In this view from Montreal near 10 AM on July 24, 2007, daylight has been turned off. Mars and the Sun are labeled. Notice that the Sun is to the east of Mars in the sky. Knowing that the default screen width is about 100°, you can see that the Sun and Mars are about 90° apart in the sky. Mars is just to the west of, and below, the Pleiades, an open cluster of stars in our galaxy, at this time. Its motion around the Sun will carry Mars across the Milky Way from our point of view during this opposition.

15. **Run time forward** and watch the motion of Mars against the background stars.

16. **Stop Time** when the Sun reappears in the western sky, in early **May 2008.**

17. Select **File > Revert** and then **Run time forward** to repeat the animation.

You can see from this long run, uninterrupted by daylight, that a long period of direct motion gives way to retrograde motion for a relatively short period between November 2007 and January 2008. You will also have noticed that the Sun was to the left (east) of Mars at the beginning of this animation and then moved eastward out of the view before returning to the view from the west, to the right of Mars. At about the mid-point in time between the disappearance of the Sun from the view and its reappearance, Mars was in retrograde motion.

> **Question 7.** How were the Sun, Earth, and Mars aligned during the time that Mars was in retrograde motion?

B. Phase and Angular Size of Mars

In the following simulation, you will view Mars from the north pole of Earth. The horizon and daylight have been removed to give an unobstructed and uninterrupted view that will allow you to observe the appearance and angular size of Mars as it moves through one synodic period, starting from conjunction with the Sun and progressing through opposition to the next conjunction.

Mars is a superior planet with an orbit larger than that of Earth. Thus, at conjunction this planet is beyond the Sun, whereas at opposition it is much closer to Earth. It therefore shows a varying size and brightness over its synodic period. Also, because it never passes between the Earth and the Sun, it will not display the full range of phases that are seen for Mercury, Venus, or the Moon. In fact, as you will see, its appearance as seen from Earth changes only slightly in the course of its orbital motion, from a full phase to a gibbous phase, with only a small region of shadow.

18. Select **Favourites > Observing Projects > Planetary Motion > Mars Phase and Size.**

Mars appears just above the Sun on October 23, 2006, at **conjunction**. By starting at this position, you can determine Mars' synodic period by measuring the time needed to return to this configuration again. This view is centered on Mars. Thus, when you advance time in intervals of 1 day, the background stars, the planets, and the Sun move in various ways, while Mars remains stationary. This mode should be maintained throughout this sequence.

In the following part of the project, you will observe Mars at various points in its motion relative to the Sun. At each point you will:

1) Record the date

2) Note the distance of Mars from the Earth

3) Observe the phase of Mars

4) Determine the angular size of Mars in the sky

Sequence 1

19. Record the **Date** of this observation from the toolbar into Data Table 1 below.

20. Open the **Info** pane for **Mars** and, under the **Position in Space** layer, find the value given for **Distance from Observer** and record this into Data Table 1 below, in the Distance column.

21. **Zoom In** on Mars to a field of view of about 1' (one arcminute). It is useful to zoom in to about the same field of view on each successive observation in order to watch the change in phase and apparent size of Mars during the following sequence.

22. Record the phase of Mars for this date in the Data Table, and if Mars shows a dark edge or limb, note which limb of Mars is dark, east (left), or west (right).

23. Under the **Other Data** layer in the **Info** pane, find the **Angular Size** of Mars in the sky and record this value into Data Table 1.

24. To ensure that the gaze is still centered on Mars, open the **Object Contextual Menu** for **Mars** and select **Centre** if necessary.

25. **Zoom Out** to a field of view 100° wide.

You can repeat the above measurement sequence twice before Mars reaches its first stationary point in order to obtain representative values of distance and angular radius of Mars as seen from Earth during this phase of direct motion.

26. **Run time forward** for four months to **February 23, 2007**. Repeat the measurements and observations described in Sequence 1 above for this date and record your results in Data Table 1.

27. **Run time forward** for another four months to **June 23, 2007**. Repeat the measurements and observations described in Sequence 1 for this date and record your results in Data Table 1.

Data Table 1. Observations of Mars Through One Synodic Period

Configuration	Date	Distance (AU)	Phase	Angular Size (")
Conjunction				
Stationary Point				
Opposition				
Stationary Point				
Conjunction				

The next three measurements should be made at stationary points and at opposition. You need to **Run time forward** until the background appears to stop moving behind Mars, indicating that Mars has reached its first stationary point. You can use single time steps to adjust Mars to this position.

28. **Run time forward** until Mars stops moving eastward against the background stars at the first stationary point. Repeat the measurements and observations described in Sequence 1 for this date and record your results in Data Table 1.

29. **Run time forward** to **December 24, 2007,** when Mars is at opposition. Repeat the measurements and observations described in Sequence 1 for this date and record your results in Data Table 1.

30. **Run time forward** until Mars reaches the second stationary point. Repeat the measurements and observations described in Sequence 1 for this date and record your results in Data Table 1.

31. Take two further sets of measurements and observations as described in Sequence 1 at about 4-month intervals.

32. Finally, **Run time forward** until Mars reaches conjunction again and is near the Sun in the sky. Repeat the measurements and observations described in Sequence 1 for this date and record your results in the Data Table above. (Hint: You might want to zoom in on Mars and the Sun to determine when Mars reaches conjunction.)

Question 8. What is the measured synodic period of Mars at this time?

Question 9. a) How much bigger does Mars appear to us at opposition compared to its appearance at the previous conjunction?

b) How much bigger does Mars appear to us at opposition compared to its appearance at the conjunction following opposition that is currently in the view?

c) Why is the apparent size of Mars at one conjunction different from its apparent size at the next conjunction?

It is instructive to re-run this time sequence at high magnification without stopping, to observe how the phase of Mars changes from full to gibbous and back again to full while its angular size also changes. The moons of Mars will be moving rapidly around the planet as this simulation proceeds. Watch particularly how the spin axis of Mars tilts to reveal more or less of the polar ice caps of this planet, an effect that produces seasons on Mars. Note also how the dark shadow edge, which makes Mars appear gibbous, moves from one side of Mars to the other as Mars moves around the Sun in its orbit.

33. Select **File > Revert.**

34. **Zoom In** to a field of view about 40" wide.

35. **Run time forward.**

36. To repeat the animation, select **Edit > Undo Time Flow** and then click **Run time forward** again.

In this simulation, Mars appears to rotate in the "wrong" direction, counter to its orbital direction. This is caused by the fact that we are advancing time in steps of 24-hour days, whereas Mars' true rotation period is about 24 $\frac{1}{2}$ hours. Thus, Mars will appear to have rotated by a little less than one rotation in every frame of the animation. This stroboscopic effect produces the illusion that Mars' rotation is "retrograde."

It is instructive to run the sequence several times to watch this impressive simulation of the varying appearance of Mars as viewed from a moving Earth and stop the time flow at the estimated time of maximum gibbous phase in order to answer the following question.

Question 10. What is your estimate of the fraction of the visible hemisphere of Mars that is still in sunlight at the maximum gibbous phase during this particular cycle?

C. Synodic Period of Mars

As mentioned in the introduction to this project, the synodic period of a planet is the time taken for the planet to travel from any particular planetary configuration as seen from the Earth to the next occurrence of the same configuration. You have already measured this period for one cycle of Mars, a superior planet, by determining the time from conjunction to the next conjunction. This is the period that is measurable from Earth.

The sidereal period of a planet is the time taken for the planet to orbit the Sun once with respect to the distant stars. This period is impossible to measure directly from Earth and must be derived from the synodic period by calculation.

The calculation of sidereal period from synodic period is simple. At this stage, we will assume that we know the sidereal period and calculate the expected synodic period of Mars. For superior planets, the equation relating the sidereal period, P, to the synodic period, S, and the orbital period of the Earth, E, is

$$\frac{1}{P} = \frac{1}{E} - \frac{1}{S}$$

In this equation, P, E, and S must be in the same units. We can rewrite this equation by adding $\frac{1}{S}$ to both sides and subtracting $\frac{1}{P}$ from both sides to give

$$\frac{1}{S} = \frac{1}{E} - \frac{1}{P}$$

Question 11. The sidereal period of Mars is $P = 687$ days, so, using $E = 365.26$ days, what value does the equation above give for the synodic period of Mars, in days?

Question 12. How does your measured time for this synodic period, from conjunction to conjunction, in Part B compare to this calculated value for the synodic period?

You will notice that the calculated value and your measured value are different! This is not experimental error or scientific uncertainty; both numbers are in fact correct values. In Part D below, you will see why these values are so different by using *Starry Night Enthusiast*™ to view the solar system from the north ecliptic pole to illustrate an important point about Mars' motion.

D. Orbital Motion of Mars from Space

You can use *Starry Night Enthusiast*™ to display the planets from a position out in space and show the actual positions of Mars and the Earth and their orbital motions during the previous sequence.

37. Select **Favourites > Observing Projects > Planetary Motion > Orbits.**

This view shows the inner solar system from 2.752 AU above the Sun in a direction toward the north ecliptic pole, perpendicular to the plane of the Earth's orbit. On this date, October 23, 2006, Mars is on the opposite side of the Sun from the Earth and is therefore in conjunction with the Sun as seen from Earth. The labeled stars, **HIP54767** and **Gamma Doradus,** serve as reference points for marking the approximate positions of Mars and the Earth in their orbits on this date, respectively.

38. Note the position of Earth with respect to the marked reference star, **Gamma Doradus**, for future reference. Also note the position of Mars with respect to the labeled reference star **HIP54767**.

39. **Run time forward** and watch Mars move with respect to the Earth, stopping the motion particularly at the positions of the planets on the date of opposition, **December 24, 2007,** and of the next conjunction, **December 5, 2008.**

40. [Optional] Select **File > Revert** and then **Run time forward** to see the animation again.

From this unique vantage point at the "north ecliptic pole," the view is directly down on the planets moving around in their respective orbits. Both orbits look quite circular, but you might notice that they are in fact elliptical because the Sun is off-center in each orbit (remember Kepler's first law, which states that the Sun is located at one focus of an elliptical orbit). You might also notice from this that Mars' orbit is more elliptical than that of the Earth. The fact that both planets follow elliptical orbits produces the discrepancy between the measured synodic period and the accepted value. In fact, this accepted value, used in the above calculation, is an average over long periods and has been determined by careful and detailed examination of the orbital paths of these planets.

Question 13. Compared to its position in its orbit at the time of Mars' conjunction on October 23, 2006, estimate how far the Earth moved around its orbit while Mars moved to opposition on December 24, 2007? (Hint: Use the star Gamma Doradus as a reference point on Earth's orbit.)

Question 14. During the interval between Mars's conjunction in October 2006 and its opposition in December 2007,

a) approximately how far has Mars moved around its orbit?

b) did Mars spend the majority of its time near perihelion or near aphelion? (Hint: The short straight line pointing away from the Sun at about the four o'clock position on Mars' orbital path marks the perihelion point of Mars' orbit.)

c) was Mars moving faster or slower than average during this part of its orbit?

You can now Run time backward and forward to explore the relative positions of Mars and the Earth over many synodic periods. You will see that some oppositions are favorable in bringing Mars closer to Earth.

41. **Stop time.**

42. Set the **Date** to the opposition of **February 15, 1995** and set the **Time Flow Rate** to **3 days.**

43. **Run time forward** and watch the relative positions of Mars and Earth, noting in particular the separations of these planets at successive oppositions. Continue to watch until the date reaches the year 2012 when you should select **Edit > Undo Time Flow.** Consider the questions at the end of this section before clicking the **Run time forward** button to watch this animation again.

You can see how the elliptical orbits of Earth and Mars can bring Mars closer to Earth at some times and not at others, depending on their positions in these orbits.

Question 15. In what year did the opposition of Mars bring it closest to the Earth?

Question 16. In what year was the distance between Mars and the Earth, when Mars was at opposition, the greatest?

E. Variations in Mars-Earth Distance at Opposition

You can now explore oppositions for 10 synodic periods of Mars, examine the variation of Mars-Earth distance and Mars angular size over these cycles of planetary motion and obtain an average synodic period for this interval.

44. Select **Favourites > Observing Projects > Planetary Motion > Synodic Periods.**

The view is again from the north pole of Earth looking at an expanded view of Mars at the opposition on June 22, 2001. The horizon and daylight have been removed in this simulation to avoid obscuration of the view as you observe Mars from Earth through 10 synodic periods.

45. Open the **Info** pane, ensure that it is displaying information on **Mars,** and expand the **Position in Space** and **Other Data** layers.

In the next sequence, you will advance time through 10 synodic periods, recording Mars' distance from Earth and angular size as seen from Earth from the **Info** pane. You will note that the **Time Flow Rate** has been set to steps of 780 days, the nominal synodic period of Mars that you calculated in Question 11 in Part C above.

46. Record the **Date** of this observation, Mars' **Distance from Observer** and **Angular Size** as given in the **Info** pane for this date in the appropriate boxes of Data Table 2, below.

47. **Step time forward** by one interval of **780 days,** the nominal synodic period of Mars.

Question 17. Judging from its phase, is Mars at opposition on this date?

Sequence 2

48. Adjust the **Date** in the toolbar until the Earth-Mars distance shown as the **Distance from Observer** in the **Position in Space** layer of the **Info** pane reaches a minimum. This can be assumed to be the position of opposition, although this is not strictly true because of the ellipticity of the orbits of Earth and Mars. For example, the date of least distance between Earth and Mars that you find for the year 2007 will not necessarily agree with the date of Mars' opposition but it will be close enough for our purposes here.

49. Record the **Date, Distance from Observer,** and **Angular Size** in Data Table 2 below.

50. **Step time forward** by one interval of **780 days.**

Data Table 2. Angular Size of Mars and Its Distance from Earth at Successive Oppositions

Opposition	Date of Opposition	Mars' Distance of from Earth (AU)	Mars' Angular Size ('')
0			
1			
2			
3			
4			
5			
6			
7			
8			
9			
10			

51. Repeat Sequence 2 above another nine times, noting the change in Mars' apparent size between successive oppositions.

From these data, you can see that the distance between Earth and Mars and the angular size of Mars as seen from Earth at successive oppositions both vary because of the elliptical orbits of these planets. You can plot graphs of these parameters to determine the form of this variation.

52. Plot a graph of the Earth-Mars distance as a function of the assigned number of the opposition.

53. Plot a second graph of Mars' angular size as a function of the assigned number of the opposition.

Question 18. Which opposition would be the most favorable for observing Mars during this time period?

Question 19. What is the form of the variation of these parameters with time, that is, are these parameters increasing, decreasing, random, periodic, or sinusoidal?

Question 20. If these variations are periodic or sinusoidal, what is the approximate period of the variation; that is, how many synodic periods elapse before the pattern starts to repeat?

Question 21. What is the average angular size of Mars at opposition during this short range of measurements?

Question 22. By what factor is the largest angular size of Mars greater than its average angular size during this time?

Question 23. What is the average distance between Mars and the Earth at opposition during this time period?

Question 24. What fraction of the average distance is the amplitude of this variation, as a percentage of the average?

The adjustments that you needed to make to move these planets to closest approach after advancing the date by the average synodic period of 780 days shows that the time period between successive oppositions varies. You can now follow a common scientific procedure that is used when data are variable, and determine an average synodic period from these data. The procedure for determining the correct synodic period for Mars is significantly more complex than this and involves a careful consideration of the orbital paths of Mars and the Earth.

54. Determine the total length of time for 10 synodic periods using the dates of the first and last oppositions from this series and divide this number of days by 10. (Note that the years 2004, 2008, 2012, 2016, and 2020 are leap years, with 366 days each.)

Question 25. What is the average synodic period for Mars during this 10-cycle sequence?

F. Variation of Synodic Period with Distance from the Sun

In this section, you can investigate how the synodic period of a planet depends on its distance from the Sun and its orbital distance compared to that of the Earth. You can use Kepler's third law and the planet's known distance from the Sun to find its sidereal period. You can then use the sidereal period to find the synodic period.

We know that a planet's sidereal period increases with increasing distance from the Sun, from a short period (88 days) for Mercury, the closest planet to the Sun, to a very long period (248 years) for Pluto, the most distant planet from the Sun. Kepler's third law gives the specific relationship,

$$P^2 = a^3 \qquad \text{Kepler's third law}$$

where P is the sidereal period in years and a is the semi-major axis of the planet's orbit in AU.

The equation in Part B,

$$\frac{1}{P} = \frac{1}{E} - \frac{1}{S}$$

applies only to **superior** planets. It can be rearranged as follows:

$$\frac{1}{S} = \frac{1}{E} - \frac{1}{P} \qquad \text{(Superior planet)}$$

As described in Box 4–1 of Freedman & Kaufmann, *Universe*, 8th Ed., the equivalent equation for **inferior** planets is

$$\frac{1}{S} = \frac{1}{P} - \frac{1}{E} \qquad \text{(Inferior planet)}$$

In each of these three equations, S (the synodic period), P (the sidereal period), and E (the Earth's period, 1 year or 365.26 days) must all be in the same units. Here, we will use years.

Data Table 3 below lists the five solar system objects Mercury, Venus, Mars, Jupiter, Uranus, and Pluto, and six hypothetical planets A, B, C, D, E, and F. The hypothetical planets are included to show how synodic period depends on distance from the Sun in regions of the solar system where there is no actual planet. Nevertheless, an object could orbit the Sun at these distances, such as an asteroid or a spacecraft.

For objects with orbital radii close to that of the Earth, to obtain meaningful sidereal and synodic periods we need to assume that the object is influenced only by the gravitational force of the Sun; that is, the object does not feel any force from the Earth, even if it passes very close to the Earth. In reality, this would not be true: The Earth's gravitational pull perturbs the orbit of any object passing close to it, sending the object into a slightly modified or even greatly modified orbit, depending on the distance of closest approach. This effect happens to asteroids and comets that pass close to any of the planets, particularly Jupiter, and has been used deliberately by NASA to save fuel by using a planetary slingshot effect to propel spacecraft to the outer planets. For the purpose of this exercise, however, we ignore this effect and look only at the effect of the Sun on the orbiting object.

The values of P, a, and S for Mercury, Venus, Mars, Jupiter, and Uranus can be found in the tables in the back of Freedman & Kaufmann, *Universe*, 8th Ed. These values are filled in for you, along with those for Pluto. For the hypothetical planets, the values of P and S need to be calculated from the specified distances from the Sun, using the method described above. For each of the hypothetical planets listed in Data Table 3, complete the steps in Sequence 3, below.

Sequence 3

55. Use Kepler's third law to calculate the sidereal period, P, for the planet and enter the value in the appropriate box in Data Table 3.

56. Use the relationship between sidereal and synodic period to calculate the synodic period of the planet, S, and enter the value in the appropriate space in Data Table 3. Be careful to distinguish between inferior planets and superior planets.

57. Repeat the previous two steps for Planets B, C, D, E, and F.

Data Table 3. Synodic and Sidereal Periods for a Variety of Real and Hypothetical Planets

Planet	a (AU)	P (years)	S (years)
Inferior Planets			
Planet A	0.10		
Mercury	0.39	0.24	0.32
Venus	0.72	0.62	1.6
Planet B	0.900		
Planet C	0.930		
Planet D	0.99990		
Superior Planets			
Planet E	1.0001		
Planet F	1.08		
Mars	1.52	1.9	2.1
Jupiter	5.20	12	1.09
Uranus	19.2	84	1.01
Pluto	39.5	248	1.00

A graph often represents the relationship between two variables with greater clarity than just a list of numbers. Use the graph template in Figure 1 below to plot the synodic period, S, along the vertical axis against the length of the semi-major axis, a, along the horizontal axis of the graph. Notice that the graph uses logarithmic scales on both axes.

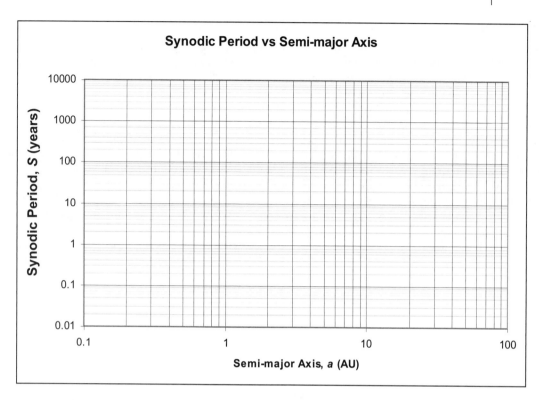

Figure 1. Graph Template for plotting Synodic Period versus Length of Semi-major Axis

Question 26. If we were to put a spacecraft into orbit around the Sun at a distance of 0.900 AU and then gradually increase the orbital distance closer and closer to 1 AU (say, 0.999 AU, 0.99999 AU, 0.999999999 AU, etc.), what would happen to the synodic period of the spacecraft? Why does this happen?

Question 27. Suppose we put the spacecraft into orbit around the Sun at a distance of 1.08 AU and then gradually decrease the orbital distance closer and closer to 1 AU (say, 1.001 AU, 1.00001 AU, etc.), what would happen to the synodic period of the spacecraft?

Question 28. Based on your answers to the previous two questions, if someone were to ask, "What is the synodic period of an object orbiting the Sun at a distance of exactly 1 AU?" what do you think the answer should be? [The mathematical process involved in obtaining this answer from the reasoning in Questions 26 and 27 is called "taking the limit of S as P approaches 1 AU"; that is, we are finding the limiting value of S as P becomes arbitrarily close to 1 AU.)

For the next two questions, look at your graph and think about the value of S shown in Data Table 3 for each planet compared to the value of P (the sidereal period of the planet) and the value of E (the sidereal period of the Earth, equal to 1 year).

Question 29. What happens to the value of S as the distance from the Sun, a, becomes very small? Why does this happen?

Question 30. What happens to the value of S as the distance from the Sun, a, becomes very large? Why does this happen?

G. Conclusions

In this project, you have followed Mars as it orbits the Sun, watching it from a moving platform, the Earth, and have seen it move from direct to retrograde motion as Earth catches up with the slower-moving Mars in its larger orbit. You have measured the change in angular size of Mars that results from this relative motion of Earth and the planet. You have observed the "real" motion of these planets from above the solar system, a position not attainable in real life. This relatively uniform motion around the Sun cannot be seen from the moving Earth.

You then examined the variability in the appearance of Mars during different oppositions, caused by the elliptical orbits of Earth and Mars, and investigated this variability over 10 successive oppositions. You then used these data to determine an average value for this variable synodic period for Mars. Finally, you explored the change in synodic period of a planet or spacecraft as its orbital distance was changed to approach that of the Earth, and when this distance was adjusted toward the limits of zero, when the spacecraft was orbiting very close to the Sun, and very large distances, when the spacecraft was moving out into the distant realms of the solar system.

Mars and its Moons   14

The planet Mars has fascinated people throughout history. Its reddish color and regular brightening over about a 2-year cycle were considered to be significant and auspicious, and this "wandering star" was regarded as a god of war by many civilizations. Later, misinterpreted descriptions of features on its surface led to the belief that Martian inhabitants had constructed canals across its surface. An extensive science fiction literature developed on these themes, and only now, after several sophisticated spacecraft have orbited and landed on Mars' surface, has this flurry of speculation given way to a more sober examination of the possibility of elementary life on this neighboring planet.

You can use *Starry Night Enthusiast*™ to look initially at the surface of Mars and then take a closer look at the two moons, Phobos and Deimos. The American astronomer, Asaph Hall, discovered these tiny moons in 1877. They were named after the mythical horses that pulled the chariot of Mars, the Roman god of war. They are irregular in shape, have average diameters comparable to that of a small city and move in almost circular orbits at relatively low altitudes above the planet's equator.

> Freedman, Geller, and Kaufmann, *Universe*, 9th Ed., discusses the Martian surface in Section 11–8 and the Martian moons in Section 11–9.

Phobos, its name meaning fear, is about twice the size of Deimos. Its period is much shorter than the rotational period of Mars. A consequence of this is that, as seen from the surface of Mars, Phobos would appear to rise in the west and move rapidly across the sky to set in the east several hours later.

Deimos, its name meaning panic, orbits at a much larger distance from the planet than Phobos and hence its orbital period is longer. In fact, Deimos' orbital motion is almost synchronous with the rotation of Mars itself and so Deimos creeps slowly across the planet's sky, taking about 3 Martian days to go from the eastern horizon to the western.

Historically, it is interesting to note that in 1726, 150 years before the discovery of these satellites of Mars, Jonathan Swift, in his satirical novel, *Gulliver's Travels*, wrote of astronomers in the land of Laputa who had discovered two satellites that revolved around Mars! Their orbital properties were described as follows:

> They have likewise discovered two lesser stars, or satellites, which revolve about Mars, whereof the innermost is distant from the center of the primary planet exactly three of its diameters, and the outermost five; the former revolves in the space of ten hours, and the latter in twenty-one and a half; so that the squares of their periodical times are very near the same proportion with the cubes of their distance from the center of Mars, which evidently shows them to be governed by the same law of gravitation that influences the other heavenly bodies.

The references to Kepler's third law of planetary motion and Newton's law of gravitation indicate that Swift, who wrote the book in the latter years of Newton's life but was not a scientist, nevertheless understood this "latest scientific theory" of gravitation. It is difficult to know what to make of his "prediction" of the existence of the moons of Mars, 150 years before their discovery! You will be able to use *Starry Night Enthusiast*™ to compare the actual parameters of the moon's orbits with these predictions.

A. Mars: Surface Features and Rotation

You can now observe Mars as it rotates and the moons as they revolve around the planet. In order to observe the planet for a reasonable period of time without interference from the horizon or daylight, the viewing location has been selected to be at the Earth's north pole in winter during a period when Mars is relatively close to Earth and at a "stationary point." At a stationary point, Mars appears to stop and reverse its direction of motion in our sky.

1. Launch *Starry Night Enthusiast*™.

2. Select **Favourites > Observing Projects > Mars and its Moons > Mars from North Pole.**

You are at the north pole at 9:00 PM on December 11, 2005 AD. You might take a moment at this stage to orient yourself at this location and look around the sky. At this time of perpetual winter at the north pole, Mars appears about 16° above the southern horizon, just to the east of a gibbous Moon. (Of course, from the north pole, all directions are south and the cardinal points of the compass around the horizon have no meaning. Nevertheless, in this simulation, they can be used to indicate directions in the sky.) Also visible in the south-east is the open cluster of stars, the Pleiades, while the spiral galaxy in Andromeda can be seen high in the sky, just to the west of south. You can identify these features of the night sky using the **Hand Tool** and move them to center-screen and zoom in on them to see their true beauty. If you scroll the view to break the lock on Mars and then **Run time forward**, you will see the stars, the Moon and Mars moving parallel to the horizon from this northernmost viewpoint as the Earth rotates around the north celestial pole.

You can now examine the surface of Mars briefly and measure its rotation period.

3. After exploring this sky, select **File > Revert.**

4. **Zoom In** to a field of view about 3' wide.

This view shows Mars with its two tiny moons and allows you to judge the scale of the moons' orbits compared to the planet's radius. You will measure these parameters later in this project.

5. **Run time forward** for a few hours to watch these moons move around the planet.

You can now examine some of Mars' surface features.

6. Select **Favourites > Observing Projects > Mars and its Moons > Mars Surface Features.**

7. Open the **Preferences** dialog window. In the **Cursor Tracking (HUD)** page, choose only **Name, Object Type** and **Surface Feature** from the **Show** list. This latter feature will allow you to identify features on Mars' surface with the **Hand Tool.**

The rotation axis of Mars at the start of this run is inclined at about 45° to the vertical on the screen. This initial view shows many dark features in the lower half of the planet below a brighter, less featured region.

8. **Run time forward** and watch Mars rotate slowly around its axis. **Stop Time** occasionally and use the **Hand Tool** to identify some of the features in the following descriptions.

As rotation proceeds, you will discover that Mars has two distinct hemispheres. Around 2:00:00 AM, a much darker region with considerable structure appears, and leads to another very dark area extending well into the northern hemisphere, an area known as **Syrtis Major Planum**. These dark regions are easily seen by 04:00:00 AM.

9. The location marker indicating Syrtis Major can be duplicated for other features on the surface of Mars. To do so, open the **Info** pane for Mars and expand the **Location Markers** layer. Here you will find a list of surface features with checkboxes that you may click to display a label and marker to indicate the location of that feature on the surface of the planet.

Around 11:30:00 AM, just as Syrtis Major is disappearing, a set of extensive canyons including Valles Marineris begins to appear that extend a significant way around the planet. By 4:00:00 PM, a line of three large volcanoes appears toward the end of these canyons.

Question 1. What is your estimate for how far this canyon system extends as a fraction of a circumference of Mars?

Question 2. What are the names of these three volcanoes?

Near 5:30:00 PM, a fourth massive volcano, Olympus Mons, has appeared, with a height about three times that of Mt. Everest and a base that extends over a distance of 600 km. The tilt of the axis of Mars is such that only a very small polar cap can be seen in the southern hemisphere at this time. We will use a different viewpoint to look at this hemisphere later in the project.

Before leaving this region of Mars, you can zoom in on the surface and examine the surface features more closely.

10. Select **File > Revert** and **Zoom In** to a field of view about **10″** wide.

11. **Run time forward** and make a note of the different kinds of surface features you see as you observe the surface of Mars rotate beneath you.

12. **Zoom In** further for a closer look at features that interest you.

Question 3. What are some of the features you see on the surface of Mars?

B. Rotation Period of Mars

In this sequence, you will measure the rotation rate of Mars.

13. Select **Favourites > Observing Projects > Mars and its Moons > Rotation of Mars.**

The view shows a close-up of Mars from the north pole of Earth. The red line across the face of Mars shows the planet's meridian facing directly toward Earth, so that this line appears to be straight at the selected time.

14. Note the **Date** and **Time** of this observation in Data Table 1 below in the row for Event #1.

15. With the **Time Flow Rate** in the toolbar at **1 minute, Run time forward** until the meridian line returns to the same position and appears straight across the face of the planet. Use the **Step time forward** and **backward** controls to adjust the time to make the meridian line precisely straight. Note the **Date** and **Time** of this observation in the row for Event #2 in Data Table 1 below.

Data Table 1. Measurement of the Rotation Rate of Mars

Event	Date	Time
Event #1		
Event #2		
Difference		

The difference obtained in Data Table 1 above actually represents the synodic rotation period of Mars as measured from a moving platform, the Earth. The required correction to allow for the movement of Earth relative to Mars is small in this case and your value can be compared to the quoted (sidereal) rotation period of this planet.

Question 4. What is the rotation period of Mars, in hours and minutes?

Question 5. Which of the other terrestrial planets has a rotation period close to that of Mars?

C. Mars' Moons from Earth

We can observe the moons of Mars from Earth and measure their orbital parameters from a view equivalent to that of an Earth-bound telescope. From these observations, we can check the "predictions" made by Swift's astronomers in *Gulliver's Travels*.

16. Select **Favourites > Observing Projects > Mars and its Moons > Mars and Orbits of Moons.**

17. **Run time forward** and observe the motions of the moons as they orbit Mars.

The moons move around the planet in the equatorial plane of Mars. The fact that the Moons move in ellipses on the screen, rather than following a straight line back and forth across Mars, shows that their orbital planes are inclined away from the Earth at this time.

As time advances, you will see that **Phobos** appears to pass behind Mars. This type of event, where an object is obscured by another object, is known as an **occultation**. Disappearance into an occultation is known as **ingress** while reappearance is known as **egress**. In order to answer the following question, you should watch very carefully the reappearances of **Phobos** from behind the planet and the appearance of **Deimos** as it moves beyond the planet on every orbit.

Question 6. Why does Phobos appear very faint or invisible when it emerges beyond the planet's limb, and what causes Deimos to appear to wink out at a certain position just beyond the planet, on every one of their orbits?

Phobos will also appear to pass in front of the planet in an event known as a **transit.** During a transit, you will note that Phobos overtakes features on the surface. In fact, it does this so quickly that, as viewed from the surface, this moon would appear to move in the same direction as the rotation of the planet, i.e., counter to the apparent sky rotation. Single one-minute time steps and a larger magnification at the appropriate time will show this relative motion more clearly. From this view, you can see how close Phobos is to the planet's surface in its orbit. In contrast, Deimos moves much more slowly in its larger orbit around the planet.

We can now determine the orbital parameters of radius and period for each moon around Mars, assuming that the orbits are closely circular, which is in fact the case.

18. Stop Time.

19. Use the **Hand Tool** to measure the angular radius of **Mars,** in arcseconds, and enter the value in **Data Table 2** below. (If in this or the following two steps the angle is more than one minute of arc, multiply the arcminutes by 60 and add this to the number of arcseconds to find the angle in seconds of arc.)

20. Measure the orbital distance of **Phobos** (the angular distance from the center of Mars to either "end" of Phobos' orbit, as seen on the screen) in seconds of arc and enter the value in **Data Table 2.**

21. In similar fashion measure the orbital distance of **Deimos** in seconds of arc, and enter the value in **Data Table 2.**

Data Table 2. Angular Radius of Mars and the Orbital Radii of its Moons

Measurement	Angular Size (")
Radius of Mars	
Orbital Radius of Phobos	
Orbital Radius of Deimos	

Question 7. By what factor is the radius of Phobos' orbit bigger than Mars' radius?

Question 8. By what factor is Deimos' orbit bigger than Mars' radius?

You can now test the "predictions" of Swift's astronomers of Laputa.

22. Calculate the orbital distances of **Phobos** and **Deimos** in kilometers, using the ratios from **Data Table 2** above and the radius of **Mars** (3400 km), and enter the values in **Data Table 3,** below.

23. Calculate the ratios of these orbital distances to the diameter of **Mars** (6800 km), and enter the values in **Data Table 3.**

Data Table 3. Orbital Radii of Mars' Moons as a Ratio of Mars' Diameter

Moon	Orbital Distance (km)	Ratio of Orbital Radius to Mar's Diameter	Swift's Ratios
Phobos			3.0
Deimos			5.0

To measure the orbital period of a moon, we need to measure the time taken for the moon to pass completely around the planet, using a reference point for timing this motion. The best reference point for Phobos is the passage of this moon across the edge of Mars' image at the beginning of a transit. (Again, this measurement will be in error because it is carried out from a moving Earth. In practice, the correction for this motion is small and can be ignored.)

24. **Run time forward** until **Phobos** is crossing a limb of Mars, adjusting the **Time** in single steps of 1 minute backward and forward to place it as close as possible to this position. Note the **Date** and **Time** in **Data Table 4** below, in the row for **Time No. 1.**

25. **Run time forward** until **Phobos** returns to the same position one orbit later and note the **Date** and **Time** in **Data Table 4** in the row for **Time No. 2.**

In the case of Deimos, this moon does not cross the planet's limb on this date as seen from the Earth. A convenient reference point is the time when it is passing over the spin axis of the planet.

26. Select **Options > Solar System > Planets-Moons.**

27. In the **Planet-Moons Options** dialog window, activate **Surface Guides** and choose only **Pole Sticks** from the sub-options. Click **OK.**

28. Use the **Time Flow** controls to position **Deimos** over the pole stick at Mars' north pole. You can use the line drawn by the **Angular Measurement Tool** from the center of Mars through this pole to align Deimos accurately. Note the **Date** and **Time** in **Data Table 4** below in the row for **Time No. 1.**

29. **Run time forward** until **Deimos** returns to this position over the north pole stick. Use single steps in time to adjust **Deimos'** position as precisely as possible and note the **Date** and **Time** in **Data Table 4** in the row for **Time No. 2.** (Hint: It might be easier to step and count the number of hours, minutes, and seconds that are stepped through to reach this second alignment to avoid the somewhat awkward subtraction of two times.)

30. The differences in these times will be the orbital periods of the moons, for comparison with Swift's predictions.

Data Table 4. Orbital Periods of Mars' Moons

	Phobos	Deimos
Time No. 1		
Time No. 2		
Difference in Times (Orbital Period)		
Swift's Predicted	10 hours	21.5 hours

Question 9. What is the orbital period of Phobos around Mars?

Question 10. How does this period compare with the rotation period of Mars?

Question 11. What is the orbital period of Deimos?

Question 12. How close were the predictions of Mars' orbital distances and periods by Swift's astronomers?

D. Mars' Moons from the Martian Surface

Because Phobos and Deimos have much shorter orbital periods than our own Moon, it is fascinating to observe their motions as seen from the surface of Mars. Visiting spacecraft to the Martian surface have observed these motions (see Figure 11-41 in Freedman and Kaufmann, *Universe*, 8th Ed.) and maybe, one day, space explorers might experience these effects in real life. Until then, we can use *Starry Night Enthusiast*™ to simulate the motions of these moons!

The brightnesses of Phobos and Deimos are very low compared to our Moon because of their small sizes. In the Martian sky, Phobos is several times brighter than Venus is in our sky, while Deimos is about as bright as Venus. Both moons move around Mars in the direction of the planet's rotation, but the orbital period of the inner moon, Phobos, is so short that it moves faster than the Martian surface. Thus, as seen by someone on the surface of Mars, it will appear to move rapidly across the Martian sky from west to east and in fact will cross this sky twice per Martian day. The orbital distance of the outer moon, Deimos, is such that it almost keeps pace with the Martian surface and is seen to move very slowly from east to west across the sky, taking more than 5 Martian days to complete a full orbit with respect to the planet's surface. In this respect, therefore, it almost mimics the synchronous communication satellites in our sky, whose orbital periods match the rotation period of the Earth accurately.

You can observe these effects in the following sequence.

31. Select **Favourites > Observing Projects > Mars and its Moons > View from Mars Surface.**

Note that the **Time** in the toolbar is specified in Universal Time. At this time and location, it is dark on Mars. If you scroll the view around the horizon, you can see the *Mars Pathfinder* rover *Sojourner* among several rocks that appear towards the NW in this simulation. If you lower your view direction, you will see the spacecraft from which this rover emerged after landing on Mars in July of 1997.

You may recognize several of the constellations in this sky, such as **Leo** just above the horizon in the southeast, **Ursa Major** high in the eastern sky, and **Gemini** high in the southern sky.

32. Toggle the display of constellations on using the **K** key to see the sky from Mars for comparison with the view from Earth.

33. Use **File > Revert** to return to the initial configuration. Select a **Time Flow Rate** of 300× and watch the sky from Mars, paying particular attention to the motion of the moons Phobos and Deimos.

As on Earth, the stars rise in the eastern (or here the south-eastern) sky and move toward the west because of the rotation of Mars. Meanwhile, **Phobos** moves quickly in the opposite direction, from west to east, and soon sets below the same horizon from which the stars are rising. You might notice that the motion of Phobos relative to the background stars is in the same direction (west to east) as that of our own Moon as seen from the surface of the Earth. However, Phobos' orbital motion is so fast that it actually moves from west to east relative to the horizon, whereas our Moon's orbital motion is so slow that the Earth's rotation causes it to drift westward along with the stars, relative to the horizon.

Deimos appears almost to hang in one spot while the stars move past it. This is an illusion created by the rotation of Mars. In fact, it is really Deimos that is moving toward the east past the stars at a rate that almost exactly keeps pace with the rotational motion of Mars. However, its orbit is not quite synchronous with Mars' rotation: If you watch closely, you may see that Deimos moves very slowly toward the west relative to the horizon. It takes just under three days to cross the Martian sky from horizon to horizon, or about 5.5 days to return to the same point in the sky again, relative to the horizon.

At about 18:45:00 UT, just as the landscape begins to brighten before sunrise, Jupiter and the Earth rise almost simultaneously above the horizon. A little over an hour later, the Sun rises into a typical Martian sky. Scattering of sunlight from the very fine dust blown into the atmosphere by frequent windstorms causes the light reddish-brown color. It is well worth re-running this spectacular simulation of a Martian sunrise!

> **Question 13.** Are any other planets visible in the sky just before sunrise in this view? If so, what are they?

It is also interesting to look at Earth from Mars as the Sun rises, with the field of view of a telescope, much as future explorers might watch the home planet from their landing site on Mars.

---

34. **Stop Time** and set the **Time** in the toolbar to **19:35:00 UT** on **December 31, 2005.**

35. Open the **Find** pane and click the menu button to the left of the listing for **Earth** and select **Magnify** from the dropdown menu.

36. Change the **Time Flow Rate** to 300× and watch the blue, white, and green Earth slowly fade as the sky brightens with sunrise on Mars.

37. To extend this look at Earth from Mars, type **Ctrl-D** to remove daylight.

38. **Stop Time** to answer the following questions.

> **Question 14.** What is the phase of the Earth at this time?

> **Question 15.** What is the angular radius of Earth when viewed from Mars at this time?

E. Mars and its Moons from Above

You can view the orbits of Mars' moons from above the plane of the ecliptic and watch the moons from a fixed reference point so that they are seen to move around the planet with their true sidereal periods.

---

39. Select **Favourites > Observing Projects > Mars and its Moons > Mars and Moons from Space.**

40. Change the **Time Flow Rate** in the toolbar to 3000× and watch the moons move around the planet.

If you look closely, you can see that Mars is also rotating in the same direction as the motion of the two moons. It is interesting to pick a Martian surface feature that is on the same side of Mars as Deimos and watch both this feature and Deimos as time passes. You might see that the relative orientation of the feature and Deimos remains almost constant, showing that Deimos' orbit is almost synchronous with Mars' rotation.

From this unique view, you can determine the sidereal periods directly by timing the passage of these moons past a convenient reference star.

---

41. Open the **Object Contextual Menu** for **Phobos** and select **Orbit** to show the orbital path of this moon. Choose a reference star very near this orbital path in the view. Place the **Hand Tool** over the chosen star and open the **Object Contextual Menu** for the star and select **Select** *Starname* to label this star for future reference.

42. Change the **Time Flow Rate** in the toolbar to **1 minute** and **Step time forward** and **Backward** to find the time when the moon is on a line drawn with the angular measurement tool from the chosen star to the center of Mars.

43. Record the **Date** and **Time** of this alignment of the moon with the star as seen from this viewpoint in the row for **Passage 1** in **Data Table 5** below.

44. Adjust the hours, minutes, and seconds in the **Time** display in the toolbar forward to move the moon around its orbit to bring it to the same position again.

45. Record the **Date** and **Time** of **Passage 2** in **Data Table 5** below.

46. Repeat the steps of this sequence for the moon Deimos.

47. Calculate the differences in the times of these events to determine the **sidereal period** for each moon.

Data Table 5. Sidereal Periods of Mars Moons

	Phobos		Deimos	
Event	Date	Time	Date	Time
Passage 2				
Passage 1				
Difference (Sidereal Period)				

Question 16. What is the sidereal period of Phobos, expressed in days?

Question 17. What is the sidereal period of Deimos, expressed in days?

From this viewpoint, you can take a closer look at the polar cap of Mars by zooming in until the field of view is about 5°. Since this northern polar cap is more easily visible when it is fully sunlit, you should change the date to October 1, 2006. This image is derived from a real Mars photograph. The radial structure at the extreme pole is caused by the image processing required for pictures taken from extreme angles by spacecraft that orbit near to the planet's equator. Nevertheless, this simulation provides a magnificent view of this remote planet and its icy polar cap.

F. Eclipse of Deimos by Phobos

Because Mars has two moons which both orbit the planet near Mars' equatorial plane, it is possible for an observer on Mars to see the inner moon, Phobos, pass in front of, or eclipse, the outer moon, Deimos. In this section, you will be able to watch a simulation of such an eclipse.

The orbits of Phobos and Deimos are inclined relative to Mars' equator by 1.08° and 1.79°, respectively. An observer close to Mars' equator would then see Phobos' and Deimos' orbits inclined relative to each other by about 0.7°, and would therefore see the orbits intersect at two points in the sky, much as we on Earth see the ecliptic and the Moon's orbit intersect at two points. If Deimos were to be near to one of these intersection points when Phobos passes by, the observer would see an eclipse of Deimos by Phobos.

48. Select **Favourites > Observing Projects > Mars and its Moons > Eclipse-Wide**.

The view is toward the east from the surface of Mars on December 31, 2005, at a location 2° south of the Martian equator. Phobos and Deimos are visible near the center of the screen, flanked by the Big Dipper on the left and Leo the Lion on the right.

49. Change the **Time Flow Rate** in the toolbar to **300×** and watch how the moons move.

As we saw earlier, Phobos moves quickly toward the east (down, in this view), opposite the apparent motion of the stars, while, over the short time represented on the screen, Deimos remains almost stationary above the horizon. At about 15:30 UT, Phobos eclipses Deimos.

50. Select **Favourites > Observing Projects > Mars and its Moons > Eclipse-Telescopic**.

The screen now shows a telescopic, or magnified, view of Deimos at a time close to the eclipse you watched in the previous sequence.

51. Change the **Time Flow Rate** in the toolbar to **30×**.

After a few seconds, Phobos passes rapidly through the area of sky visible on the screen, eclipsing Deimos as it does so.

52. Use **File > Revert** to return to the initial view.
53. **Run time forward** and then **Stop Time** when Phobos is near the upper edge of the screen.
54. Open the **Object Contextual Menu** over each moon in turn and select **Orbit** to display the orbital paths of these moons.
55. Change the **Time Flow Rate** in the toolbar to **1×** and then click the **Run time forward** button to watch the eclipse again in real time. Continue to watch the view after the eclipse and notice the point at which the orbital paths of the two moons intersect. You can then select **Edit > Undo Time Flow** and then **Run time forward** to watch this eclipse again.

Question 18. How long, in seconds, does this eclipse of Deimos by Phobos last? (For the beginning of the eclipse, note the time when Phobos first contacts Deimos and for the end of the eclipse, note the time when Deimos is again completely visible in the view.)

You can see that Deimos is indeed near an intersection point of the two orbits. You can also see that, because Phobos appears much larger than Deimos in the sky and because the orbits are inclined at such a small angle relative to each other, Deimos can actually be some distance away from the intersection point and still be eclipsed by Phobos. *Starry Night Enthusiast*™ shows these moons to be spherical in this simulation but they are actually somewhat irregular and cratered.

G. Conclusions

The appearance and rotation of Mars and the revolution of its two small moons have been examined from various novel viewpoints, including the surface of the planet itself. These observations were used to measure the radii and periods of the orbits of the moons. These values were then compared to the fictitious "predictions" made by Jonathan Swift in his novel, *Gulliver's Travels*, 150 years before the moons were actually discovered! You have also viewed the eclipse of one moon by the other, a phenomenon impossible on Earth because our planet has only one moon.

Kepler's Third Law 15

Several thousand years ago, Greek philosophers followed logic and the principles of scientific investigation to develop models that described the motion of the "wandering stars" in the sky, the so-called planets. These geocentric models were based on the understandable belief that the Earth was stationary. These models required many arbitrary assumptions and only over relatively short periods were they able to predict the future motion of the planets with reasonable accuracy. Over longer periods, the geocentric models required constant revision and became more and more complex.

Copernicus, in the sixteenth century, was able to demonstrate that a model in which the planets orbited the Sun could explain these motions more simply. In his heliocentric model, he was able to show that the apparent complexity of the planetary motions arose from the fact that we were watching these planets from a moving Earth that also orbited the Sun.

By 1609, Johannes Kepler had developed this idea and had demonstrated that a simple model, in which the planets were required to obey only three laws, could describe the motion of the planets around the Sun:

1) The orbit of a planet about the Sun is an ellipse with the Sun at one focus.

2) The line joining a planet to the Sun sweeps out equal areas in equal intervals of time.

3) The square of the sidereal period, P, of a planet's orbit around the Sun is directly proportional to the cube of its semimajor axis, a; that is,

$$P^2 = \text{constant} \times a^3 \qquad \text{(Equation 1)}$$

where the constant is the same for all planets in the solar system.

For an elliptical orbit, the semimajor axis, a, is half the longest diameter of the ellipse; that is, the distance from its center to one end. The sidereal period, P, is the length of time the planet takes to complete one orbit measured relative to the distant stars.

If we measure P in years and a in AU, then we can write Equation (1) more simply as

$$P^2 = a^3 \qquad \text{(Equation 2)}$$

Equation 1 is actually true for any object in orbit around any other object, provided that the mass of the orbiting object is much smaller than the mass of the object it is orbiting. Of course, the constant of proportionality will be different for different situations. For example, the satellites (moons) of all of the major planets obey Kepler's laws. In this project, after a brief glance at Jupiter itself, you can use *Starry Night Enthusiast*™ to investigate the motions of Jupiter's Galilean satellites. You can measure the semimajor axes and periods of each of these four moons and use these measurements to verify Kepler's third law. Then, with a little extra calculation, you can use your measurements to find the mass and mean density of Jupiter.

> Kepler's laws are discussed in Section 4–4 of Freedman, Geller, and Kaufmann, *Universe*, 9th Ed.

A. Jupiter's Rotation Period

It is interesting to spend a short time examining Jupiter at close quarters before going on to explore the Galilean moons and to use their motions to verify Kepler's third law. In particular, it is interesting to measure the rotation period of this large and massive planet, particularly when compared to that of the much smaller Earth. Furthermore, we can see and measure the effect of the rapid rotation on this largely fluid planet.

1. Launch *Starry Night Enthusiast*™.
2. Select **Favourites > Observing Projects > Kepler's Third Law > Jupiter.**

The view is from the north pole of Earth at 11:30 PM on February 22, 2001 AD. The two poles at the north and south of the planet show the spin axis of Jupiter. A meridian line is depicted on the planet and will be used to measure the rotation period. You can see a great deal of detail in this 1 arcminute field of view.

3. **Run time forward** with a **Time Flow Rate** of **1 minute.**

Jupiter has no solid surface; you are seeing the tops of clouds above a very deep atmosphere. The obvious turbulence and structure in this atmosphere, including the Great Red Spot just appearing on the east side of Jupiter, is caused by differential rotation, where different latitudes of the planet rotate at different rates. Unfortunately, the present simulation does not include this differential rotation, and you can measure only the average rotation period for Jupiter by timing the motion of the defined meridian.

You can see that Jupiter rotates rapidly, bringing the Red Spot to face Earth within an hour or so. Meanwhile, Io, the innermost Galilean moon, appears to the east of the planet at just after midnight and moves rapidly toward transit of the planet, followed shortly by its shadow on the cloud-tops.

There is one interesting puzzle about this kind of shadow on Jupiter. The shaded region, in infrared light, appears hotter than the surrounding atmosphere, contrary to expectation. The reason for this is probably that the momentary blocking of sunlight causes clouds in Jupiter's atmosphere to dissipate, allowing infrared detectors to see more deeply into warmer regions of the planet's atmosphere. A related effect is the observation that the belts, relatively cloud-free regions around Jupiter that appear dark in visible images, actually appear bright in infrared images, suggesting that they are warmer than the adjacent optically bright (but infrared-dark) zones that are probably high cloud-tops. You can search for the shadows of other moons, particularly Io and Ganymede, as they transit the planet.

4. Select **File > Revert** and then **Run time forward** and **backward** and also use the **Step Time** controls in 1-minute intervals to answer the following questions.

Question 1. At what time is the Red Spot directly facing the Earth during this rotation of Jupiter?

Question 2. When does Io begin its transit across the face of Jupiter during this rotation period?

Question 3. When does Io's shadow begin to cross the planet's face?

Question 4. From the appearance of this shadow following Io across Jupiter, from which direction is the Sun shining at this time, the left or right of the Earth or directly behind the Earth?

You can now go ahead and measure the average rotation period of Jupiter

5. Select **File > Revert** to return to the original view.

6. Adjust the **Time** so that the meridian is directly in line with the north and south poles and note this **Time**.

7. *Alternative 1*: **Run time forward** until this meridian returns to the view and then **Step Time** in intervals of **1 minute** to precisely align the meridian with the poles again and note this **Time**. The difference between this time and that of the previous step will be the rotational period of Jupiter.

   Alternative 2: An alternative method for accomplishing this step is to highlight and then increment, using the + key, first the **hour** field and then the **minute** field in the **Time** display in the toolbar to advance time in hours and minutes. Count the number of hours and minutes as you advance time with each press of the key. If you overshoot, press the − key and subtract one from your count. The total count of hours and minutes will be the rotation period of Jupiter, and you will not need to calculate the difference between the two times.

Question 5. What is the average rotation period of Jupiter?

A fluid object rotating this rapidly will be distorted and its equatorial diameter will be larger than its polar diameter. You can measure the effect of this rotation on the shape of this fluid planet by measuring the equatorial and polar radii. To do so, you will want to see the planet in its full phase as seen from Earth.

8. Change the **Date** in the Toolbar to November 29, 2000 AD.

9. Use the **Angular Measurement Tool** to measure the angular size of the equatorial and polar radii of Jupiter as seen from Earth on this date. To find the angular size of the equatorial radius, measure from the center of Jupiter to the east or west limb (edge) of the planet. To measure the angular size of the polar radius of Jupiter, measure from the center of the planet to its north or south pole.

Question 6. What is the angle subtended at the Earth by the equatorial radius of Jupiter?

Question 7. What is the ratio of the equatorial radius to the polar radius of Jupiter?

B. Jupiter's Moons

Before beginning the detailed measurement of the orbital properties of its moons, it is worth taking the opportunity to watch these moons as they orbit the planet and explore several events that occur around Jupiter. You have already watched a transit of a moon and its associated shadow. A related phenomenon is the occultation of a moon by the planet.

10. Open the **Preferences** dialog window and select the **Cursor Tracking (HUD)** page. Select **Name** and **Object Type** from the **Show** list. Then close the dialog window.

11. Select **Favourites > Observing Projects > Kepler's Third Law > Galilean Moons**.

The start time and date of observation is midnight, February 23, 2001 AD, and the observing site is again the north pole of the Earth. The field of view is 15 arcminutes wide (half the diameter of the Full Moon in our sky), and shows Jupiter with the four Galilean satellites near it. The **Hand Tool** can be used to identify each moon in turn.

> 12. **Run time forward** with the **Time Flow Rate** set at **10 minutes** and watch the Galilean moons orbit Jupiter. To help you to answer the following questions, you can display labels and the orbital paths of these moons by opening the **Find** pane and clicking the boxes on either side of the names of the Galilean moons, **Io, Europa, Ganymede,** and **Callisto,** listed under **Jupiter.**

> **Question 8.** Which of the Galilean satellites moves in the smallest orbit and completes its orbit in the shortest time?

> **Question 9.** Which of the Galilean satellites moves in the largest orbit and completes its orbit in the longest time?

The four satellites move around Jupiter in nearly circular orbits in the plane of Jupiter's equator. Jupiter's equatorial plane almost coincides with the ecliptic (that is, with the Earth's orbital plane) and so we see these orbits almost edge-on. Thus, we see each satellite apparently moving back and forth past Jupiter in almost a straight line, the nearest ones to the planet moving fastest, as expected from Kepler's laws. At this time and with this geometry, three of the four moons undergo regular occultations, where the moon passes behind the planet. Disappearance is known as ingress, while reappearance is known as egress. These three moons also transit, or pass in front of, the planet.

Several features of these motions are worthy of note. Two of these moons suddenly reappear after being occulted by Jupiter, but at a distance to the east of Jupiter not at the planet's limb. Furthermore, a third moon is occulted and reappears at Jupiter's east limb only to wink out again for a short period of time. This effect can be examined in more detail by zooming in on Jupiter.

> 13. Select **Favourites > Observing Projects > Kepler's Third Law > Ganymede.**

Ganymede, the third most distant of the Galilean moons from Jupiter, can be seen just before ingress.

> 14. **Run time forward** and observe the unfolding sequence of events.

Watch Ganymede particularly as it moves ahead in its orbit at around 4:30 PM, after it has been occulted by and has reappeared from behind Jupiter.

Europa, the second of the Galilean moons from Jupiter, appears from the left and transits the planet, starting at about 7:35 PM on February 23, 2001.

Io, the innermost Galilean moon appears on the right of the screen and goes into occultation at about 11:40:00 PM. Watch this moon particularly as it reappears from behind the planet.

> **Question 10.** In view of the geometry of Jupiter and its moons, why do these moons appear somewhat dimmer at certain times following egress from occultation?

There is one more event that you can witness with this simulation, namely the occultation of a star by Jupiter. Observations of the star and its spectrum during such an event, particularly as Jupiter's atmospheric layers occult the star, have been used to determine the chemical make-up and structure of the planet's atmosphere.

15. Select **Favourites > Observing Projects > Kepler's Third Law > Io**.

16. **Run time forward** and watch both Io and its shadow traverse the planet while Jupiter passes in front of the star, **TYC1258-377-1**. In this view, the gaze is locked onto the star and you see Jupiter moving and rotating, much as you would through a telescope.

By coincidence, at the beginning of this simulation, Io's orbital motion toward the west almost exactly compensates for Jupiter's orbital motion toward the east, so that Io appears to be stationary in our sky. If you continue to watch, Io will gradually start to move eastward as it approaches the western "end" of its orbit as viewed from the Earth. In this simulation, the star image appears larger than it would appear in the real sky. Again, you see the shadow of Io following the moon across the face of the planet, thereby indicating the direction of the Sun.

C. Motions of the Galilean Satellites of Jupiter

You can verify Kepler's third law, the relationship between semimajor axis and orbital period, using further observations of the detailed motions of the Galilean moons of Jupiter.

17. Select **Favourites > Observing Projects > Kepler's Third Law > Kepler 3**.

This view of Jupiter and its moons is again from the north pole of Earth to avoid interference with the horizon. On this date in late November 2000 AD, Jupiter is close to opposition and is well above the horizon from this site.

The orbits of the four Galilean moons are almost precisely circular (with eccentricity 0.01 or less), so the length of the semimajor axis is simply the radius of the orbit, or half the diameter. To measure the length of the semimajor axis of any moon's orbit, therefore, we can measure the length of the line from the center of Jupiter to the moon when this moon is at its farthest position from Jupiter, as seen on the screen.

To measure the orbital period of a particular moon, we can adjust the time to place the moon at a specific and repeatable position with respect to Jupiter and simply count the number of days, hours, and minutes that are required to bring the moon back to this position. In the case of Io, Europa, and Ganymede, we can use the time of ingress of the moon behind the western edge of Jupiter. At the time of the animation in this section, Jupiter does not occult Callisto. Thus, we can use the time when Callisto is directly above the north pole of Jupiter, a position that is easy to define with the extended north pole axis that can be introduced by *Starry Night Enthusiast*™.

Sequence 1 below details the steps involved in obtaining this data for a selected moon.

Sequence 1

18. Select a Galilean moon and use the procedure outlined in the following steps to determine the semimajor axis and revolutionary period of its orbit.

19. Open the **Find** pane, expand the list of moons under **Jupiter,** and click the box to the right of the selected moon to display its orbit.

20. If necessary, adjust the **Zoom** factor of the view so that the moon's orbital path almost fills the view.

21. Manipulate the **Time Flow** controls to place the selected moon at one extreme of its orbit.

22. Measure the angular distance between the center of Jupiter and the selected moon. Record this distance in arcminutes and arcseconds under the column labeled **Measured Orbital Radius** in Data Table 1 below.

23. Convert this angle into arcseconds (1' = 60'') and record this number under the column labeled **Semimajor Axis, *a*,** in Data Table 1.

24. Use the **Time Flow** controls to move the moon to a position to the west (right) of the planet on the far side of its orbit.

25. **Zoom In** until Jupiter almost fills the field of view and adjust the **Time** until the moon is at ingress, with half of the moon occulted by the planet's edge. (In the case of Callisto, move the moon to align it with the north end of the Jupiter's spin axis, shown by a colored pole.)

26. **Zoom out** until you can see most of the moon's orbital path.

27. Use the keyboard shortcuts **D** and **H**, respectively, to increment first the day and then the hour and step and count the number of days and then hours to move the moon round its orbit and return it close to the western edge of Jupiter. If you overshoot, decrement the day or hour by pressing the **Shift** key as you type the **D** or **H** key, respectively.

28. **Zoom In** again to watch this approach carefully. Use the **T** key to step and count the extra minutes needed to move the moon to the precise position of ingress, and note the days, hours, and minutes needed to complete this orbit in Data Table 1 under the column labeled **Measured Period**. (In the case of Callisto, return the moon to its original position above Jupiter's north pole.)

29. Convert the **Measured Period** to decimal days by dividing the number of minutes in the measured period by 60, adding this fraction to the number of hours, dividing this sum by 24 to convert it to a fraction of a day, and finally add this fraction to the number of days in the **Measured Period**. Record this number under the column labeled **Period, *P*,** in Data Table 1.

30. Select **File > Revert** and repeat this sequence for the remaining three moons.

Data Table 1. Semimajor Axes and Orbital Periods of the Galilean Moons of Jupiter

Moon	Measured Orbital Radius		Semimajor Axis, *a*	Measured Period			Period, *P*
	'	''	''	Days	Hours	Minutes	Days
Io							
Europa							
Ganymede							
Callisto							

Before using your data to verify Kepler's third law, the following questions outline an interesting relationship between the periods of these moons.

Question 11. What number do you get if you divide the orbital period of Europa by the orbital period of Io? Is this number close to an integer? Which integer?

Question 12. What number do you get if you divide the orbital period of Ganymede by the orbital period of Io? Is this number close to an integer? Which integer?

Question 13. What number do you get if you divide the orbital period of Ganymede by the orbital period of Europa? Is this number close to an integer? Which integer?

Question 14. From your answers to the above questions, is there any clear synchronicity between the orbital periods of Io, Europa, and Ganymede?

D. Verifying Kepler's Third Law

Equation 1 represents Kepler's third law for objects orbiting a much more massive object. If the orbits are circular, then P in Equation 1 is the orbital period and a is the orbital radius. If we write the constant in Equation 1 as k, then Kepler's third law becomes

$$P^2 = k \times a^3$$

where the numerical value of k depends on on the units that are used for P and a. We can rearrange this equation as follows:

$$k = \frac{P^2}{a^3}$$

(Equation 3)

Therefore, you can verify Kepler's third law by demonstrating that the value of k is the same for each of the Galilean moons. Data Table 2 below allows you to calculate the value of k for each of these moons.

> 31. Copy the values of **Period, P,** in days and **Semimajor Axis, a,** in arcseconds from Data Table 1 into Data Table 2 below.
> 32. Use Data Table 2 to calculate P^2, a^3, and the value of k for each moon.

Data Table 2. Verification of Kepler's Third Law

Moon	P (Days)	P^2 (Days)^2	a (")	a^3 (arcseconds)^3	$k = \dfrac{P^2}{a^3}$
Io					
Europa					
Ganymede					
Callisto					

If, from your data, k has the same value for all four moons, then you have verified Kepler's law as it is applied to the Galilean moons of Jupiter. For the data used here, the units of k are $\dfrac{\text{(days)}^2}{\text{(arcseconds)}^3}$.

Question 15. According to your data, do the moons of Jupiter obey Kepler's law?

Question 16. Suppose space scientists wanted to place a spacecraft in orbit around Jupiter to move synchronously with the surface of the planet, to study the Red Spot in detail, for example. Use the average rotational period of Jupiter measured in Question 5 of Section A and the above equations to calculate value of a, the angular radius of the required orbit for this spacecraft if measured from Earth.

Question 17. Is this orbit possible; that is, is this orbital radius greater than the equatorial radius of Jupiter?

Question 18. If the equatorial radius of Jupiter is 71,500 km, use the equatorial angular radius measured in Question 6 of Section A to calculate the actual orbital radius for this spacecraft in kilometers.

E. Measuring the Mass of Jupiter

The measurements of the orbital motions of moons around a planet offer the best method for measuring the mass of the planet itself using a version of Kepler's third law derived from Newton's gravitational law. Indeed, the measurement of the gravitational influence of one object on another, such as one component of a binary star system or a galaxy moving close to a second galaxy, is the standard method for determining masses of these objects.

For the special case of an object moving in a circular orbit around an object whose mass is much larger than itself, Kepler's third law becomes

$$R^3 = \left[\frac{GM}{4\pi^2} \right] \times P^2 \qquad \text{(Equation 4)}$$

where R is the radius of the satellite's circular orbit, P is the satellite's orbital period in seconds, M is the mass of the central object (in the present case, Jupiter) in kg, and G is a universal gravitational constant, applicable throughout the universe. Since the Galilean moons are in circular orbits and the mass of each Galilean moon is less than 1/1000 the mass of Jupiter, the required conditions are fulfilled for the application of this equation to the motion of these objects.

Multiplying both sides of Equation 4 by $4\pi^2$ and dividing both sides by GP^2, then rearranging, gives

$$M = 4\pi^2 \times \frac{R^3}{GP^2} \qquad \text{(Equation 5)}$$

You can derive the mass of Jupiter using Equation 5 and your measurements of the orbital parameters of one or other of the Galilean moons. You have first to translate the values of R and P into units of meters and seconds, respectively. G is the gravitational constant with a value of 6.673×10^{-11} N $\cdot$ m^2 $\cdot$ kg^{-2}. The mass of Jupiter, M, can then be determined, in kilograms.

You can use any of the satellites to calculate Jupiter's mass, but it is probably best to choose Callisto. This is because Callisto has the largest values of R and P, so the uncertainties in measurement are the smallest, relative to R and P themselves.

33. Assuming that Callisto's orbit is circular, you can use the value measured as the Semimajor axis, a, of Callisto's orbit as recorded in Data Table 1 as the value for R. You can use the angular equatorial radius of Jupiter measured for Question 6 and the fact that Jupiter's equatorial radius is 7.15×10^7 meters to calculate the orbital radius of Callisto in meters (use simple proportionality to calculate this). You will use this value for R in Equation 5 to calculate the mass of Jupiter.

34. You measured the orbital period of Callisto in days in Data Table 1. Multiply this orbital period of Callisto in days from Data Table 1 by $24 \times 60 \times 60$ (= 86,400) to obtain this in seconds, for use as P in Equation 5.

35. Use Equation 5 to calculate the mass of Jupiter in kilograms.

Question 19. What is the mass of Jupiter in kilograms?

Question 20. How does this mass compare to the mass of the Earth, which is 5.974×10^{24} kg?

F. Measuring the Mean Density of Jupiter

The above calculations allow you to calculate the average density of this massive body, in kg/m^3, and compare this with the density of common materials such as, for example, water, with a density of 1000 kg/m^3.

The average density, ρ, of a planet is equal to the mass of the planet divided by its volume:

$$\rho = \frac{M}{V} = \frac{M}{\left[\frac{4\pi R^3}{3}\right]} = \frac{3M}{4\pi R^3} \qquad \text{(Equation 6)}$$

where M and R are the planet's mass and radius, respectively, and V is the volume of a sphere. Here we are assuming that Jupiter is close enough to a sphere in shape that we can neglect its rotational flattening.

36. Use Equation 6 to calculate the average density of Jupiter, using the calculated mass and the given radius of Jupiter.

Question 21. What is the density of Jupiter in kg/m³?

Question 22. How does your result compare to the density of a) water (1000 kg/m³), b) the density of rock (about 2800 kg/m³), and c) the mean density of the Earth (5500 kg/m³)?

Question 23. What does your answer to the last question suggest about the bulk composition of Jupiter?

G. Conclusions

You have examined Jupiter's surface, measured its average rotation period, and watched moons and their shadows transit it's face and become occulted behind the planet. Furthermore, you have used measurements of the motions of these moons to verify Kepler's third law of planetary motion, and you have carried out a classic experiment to measure the mass of Jupiter from these measurements. These results have allowed you to compare the average density of this massive planet with the density of common materials and speculate about its composition.

Comets 16

Comets are perhaps the most enigmatic of objects in our solar system. Their appearance in the sky has invoked both fear and admiration in many civilizations. The tail of a bright comet can stretch across half the visible sky and be as bright as most stars, if only for a few days. Nevertheless, there is little substance to these ephemeral visitors.

The comets that appear in our sky are either regular visitors to the inner solar system, moving in stable orbits around the Sun, or they are first-time visitors from the distant realms of the solar system. This latter group may be in very long elliptical orbits that carry them far from the Sun with very long orbital periods. Those with periods shorter than 20 years originate in the Kuiper belt, a flat region of icy objects near the plane of the ecliptic, extending from about 30 AU (just beyond the orbit of Neptune) to 50 AU from the Sun. Those with longer periods, from 20 years to about 30 million years, come from the Oort cloud, a spherical region of icy objects extending from 50 to more than 50,000 AU from the Sun. Whereas Kuiper belt objects move in roughly circular orbits and probably formed where we see them now (except when collisions send fragments plummeting into the inner solar system as comets), the objects in the Oort cloud formed in the region of the Jovian planets and were flung out to great distances by gravitational encounters with the Jovian planets billions of years ago.

> Comets are discussed in Section 15–7 and Section 15–8 of Freedman, Geller, and Kaufmann, *Universe*, 9th Ed.

The solid part of a typical comet, the comet's **nucleus**, consists of a ball of ice, dust, and rock a few kilometers across. As this insignificant snowball comes close to the Sun, ices in its outer layers evaporate to release atoms, molecules, and dust grains from within the ice into a large cloud surrounding the nucleus. Much of this thin cloud of released material is blown outward from the Sun into a long tail by the action of the solar wind and sunlight. These components produce different parts of the comet's tail. The **gas tail** is composed of the lighter atoms and molecules that are blown directly outward in a radial direction from the Sun by the solar wind and shows fluting and structure in response to variations in this wind and its associated magnetic field. Within a certain distance of the Sun, these atoms and molecules are excited by solar UV radiation and emit light whose spectrum is intrinsic to the specific components. The **dust tail** is composed of the much more massive dust grains that are pushed more gently outward from the comet's orbit, mostly by radiation pressure from sunlight. This component of the comet's tail shows a curved shape with little or no structure

> Figure 15–24 in Freedman, Geller, and Kaufmann, *Universe*, 9th Ed., illustrates the two parts of a comet's tail.

and is seen by scattered sunlight except when very close to the Sun, where metallic atoms are induced to emit their characteristic emission spectra by excitation by intense solar radiation. In this way, even though a very small amount of matter is involved, a comet can stretch itself across a significant region of our sky and become disproportionately bright during its brief visit to the Sun's vicinity. For example, the tail of Comet Hyakutake (1996) was at least 4 AU in length.

In this project, you will look at several comets and watch in particular the development and alignment of the celebrated and much-studied Comet Halley, named for Edmund Halley in recognition of his prediction of its return in 1758. This prediction was shown to be correct when the comet was "discovered" on Christmas night of 1758. Regrettably, this observation occurred long after Halley's death! Halley's insight that the appearance of a major comet in 1531, 1607 and 1682 was in fact a returning object and the detailed calculation of its orbit using Newton's newly developed theory of gravitation was a major milestone in the development of a rational explanation for the behavior of objects in the solar system. Comet Halley has been observed in every one of its returns to the vicinity of the Sun on a 76-year cycle since 239 BC.

The simulations in this project will show the general appearance and behavior of Comet Halley. You will observe the growth, decay, and direction of its tail during the comet's last appearance in 1986, although the detailed and changing structure of this tail under the influence of the variable solar wind is difficult to reproduce in simulation. There are several superb images of comets and their tails in Chapter 15 of Freedman, Geller, and Kaufmann's book, *Universe*, 9th Edition, particularly of Comet Hale-Bopp in the introductory image of this chapter and in Fig. 15-24 and Comet Hyakutake in Figure 15-19. The images of the nuclei of two comets, Halley and Tempel, taken by spacecraft cameras, are shown in Figure 15-21.

You will also be able to verify Kepler's second and third laws of planetary motion with Comet Halley. The second law, originally stated by Kepler in the form, "As an object orbits the Sun, the line joining planet to Sun will sweep out equal areas in equal times", leads to the conclusion that the speed of any object moving under the influence of the Sun will be higher, the closer the object is to the Sun. Comet Halley moves in a long elliptical orbit, and its distance from the Sun, and therefore its orbital speed, changes over a wide range. The motion of this spectacular object thus provides a good general test of this law.

A. Comet Tails

We will first observe Comet Halley from the Earth's surface. The chosen site is Caracas, Venezuela, at latitude 10° N.

1. Launch *Starry Night Enthusiast*™ and select **Favourites > Observing Projects > Comets > Halley**.

2. Open the **Preferences** dialog and set the **Cursor Tracking (HUD)** options to show **Distance from Observer, Name**, and **Object Type**. Then close the **Preferences** dialog window.

The view is toward the southeast from Caracas at 4:00:00 AM on March 7, 1986. Comet Halley can be seen just above the horizon in the SE, with its tail extended upward.

3. Change the **Time Flow Rate** in the toolbar to **300×** and observe Comet Halley until the increasing brightness of daylight hides it from view.

As time advances, you will see Comet Halley rise majestically from the horizon, along with the star background. The blue color of its tail is caused by scattering of sunlight from small dust particles in the comet's tail. We can infer from the color of this scattered light that the typical size of these dust particles is equivalent to the smoke particles from a campfire. As sunrise approaches, the comet's tail slowly fades into the morning twilight.

4. **Stop Time.**

5. Set the **Time** in the toolbar to **6:45:00 AM** on **March 7, 1986.**

6. In order to see the comet more clearly at this time, select **View > Hide Daylight.**

Question 1. In which direction does the tail of Comet Halley point relative to the Sun at this time in March?

It is interesting to observe two further comets that appeared in the sky in the decade or so after the last appearance of Comet Halley and compare their behavior with this comet.

7. Select **Favourites > Observing Projects > Comets > Three Comets.**

The view is again from Caracas, but the date has been set back to January 29, 1986, at 7:30:00 AM. The gaze is locked on the Sun, centered in the view. Daylight has been removed, so that the stars and Comet Halley are visible in the morning sky.

Question 2. In which direction does the tail of Comet Halley point on this date, relative to the Sun at this time in January?

8. With the **Time Flow Rate** in the toolbar set to **1 day, Step time forward** and observe the orientation and length of the tail of Comet Halley until it moves out of the view.

Question 3. Relative to the Sun, throughout the time span of the previous step, how is the tail of Comet Halley oriented?

9. Change the **Date** in the toolbar to **April 21, 1996.**

The view shows the same morning sky from Caracas with daylight removed and the gaze centered on the Sun, but it is now about 10 years later and a different comet appears in the sky.

10. **Step time forward** at 1-day intervals and observe the orientation and length of the tail of this comet. Use the **Hand Tool** to identify this comet.

Question 4. Which comet visited the inner solar system in April 1996?

11. Change the **Date** in the toolbar to **February 25, 1997.**
12. **Zoom Out** to a field of view **100°** wide.

You are still in Caracas and it is still 7:30 AM, but it is now 1997, at a time when another famous comet, Hale-Bopp, made a spectacular appearance.

13. Observe the orientation of the tail of this comet as you **Step time forward** in 1-day intervals until Comet Hale-Bopp moves below the horizon.

Question 5. Is the orientation of the tail of Comet Halley relative to the Sun a unique trait of this comet?

Question 6. Despite a sample of only three comets, what conclusion can you draw from your observations about the direction of a comet's tail relative to the Sun?

B. Observations of Comet Halley from Earth

In this section we return to our examination of Comet Halley. It is possible to watch the movement of this comet over a longer period against the background stars by stepping time in intervals of sidereal days. This will keep the stars in exactly the same position in your view, night by night.

14. Select **Favourites > Observing Projects > Comets > Halley.**

15. With the **Time Flow Rate** in the toolbar set at **1 sidereal day**, observe the comet as you use the **Step time forward** button to move forward in time to **April 13, 1986.**

In this sequence, it is obvious that the direction of the comet's tail is not controlled by motion of the comet itself because the tail precedes the comet nucleus on this section of its orbit. In fact, the tail is controlled by the direction of sunlight and the prevailing solar wind.

You can watch the majestic sweep of the comet as it moves away from the Sun after this close approach.

16. Use the **Hand Tool** to adjust the gaze direction so that Comet Halley is near the left edge of the screen on **April 13, 1986.**

17. Use the **Single Step Forward** button to advance time until you can no longer see the comet's tail.

Question 7. Approximately when does Comet Halley's tail disappear from view?

You can observe this comet's approach to the Sun and its recession into space again without obscuration by the scattered blue light of the daylight sky.

18. Select **Favourites > Observing Projects > Comets > Daylight Removed.**

The view is of the SE sky from Caracas at 10:00 AM on January 7, 1986, with daylight removed. At this time, the Sun is about 35° above the horizon, along with the planets Venus and Mercury. The planet Jupiter can also be seen above the eastern horizon, close to the comet.

19. Identify the Sun, Venus, Mercury, and Jupiter with the **Hand Tool.**

20. Use the "U" key to **Step time forward** in steps of **1 sidereal day** to **April 8, 1986.**

As time progresses in time steps of 1 sidereal day, the Sun moves eastward toward its sunrise position, accompanied by its entourage of planets. Meanwhile, the comet rises as it approaches the Sun and we see the foreshortened tail pointing away from the Sun at the time of the comet's closest approach to the Sun in February.

By the end of March and early April, when the comet has moved to its position of closest approach to Earth, we see a spectacular side-on view of its tail. As we shall see from other viewpoints, this is not the time of maximum intrinsic tail length. This will be when the comet is closest to the energy source that evaporates the ice to produce its tail, the Sun. It is also interesting to note that the tail changes materially day by day such that, were you to be watching the comet in real life, you would be viewing a new tail every day! For all its glorious appearance, the amount of material within this tail is miniscule and could probably be packed into a small suitcase. Every atom, molecule or grain of dust is either emitting light or scattering sunlight toward us to produce this great sight.

After this time period, the tail becomes progressively smaller. Apart from the apparent reduction in length caused by foreshortening as we see the tail from an end-on view, there is also a real reduction in the length of the tail as the comet moves away from the Sun. Solar heat falling on the comet's nucleus is reduced as it moves away from the Sun and less of this nucleus is being evaporated to form the tail.

21. Use the **Hand Tool** to adjust the **Gaze** to the **SW**.

22. Open the **Options** pane and under the **Local View** layer, click the checkbox beside **Local Horizon** to remove the horizon from the view.

23. **Run time forward** and observe the comet as it moves away from Earth and into space until, by August 1986, it is almost invisible.

After this brief visit to our vicinity, Comet Halley will return again only after about another 75 years. Most of this time will be spent as a cold, icy nucleus, unaccompanied by a tail and invisible from Earth.

This sequence has given you a panoramic view of the full sweep of the comet and its tail during the close approach to the Sun, projected against a fixed sky without the hindrance of daylight, an opportunity not possible in real life! You can repeat this sequence and use the **Info** facility to monitor the distance of Comet Halley from both the Earth and the Sun to determine the dates of closest approach to each of these objects.

24. Select **File > Revert**.

25. Open the **Object Contextual Menu** for **Comet Halley** and select **Show Info**.

26. In the **Info** pane, expand the **Position in Space** layer. Note the values for **Distance from Observer** and **Distance from Sun** and watch them as you **Step time forward** in intervals of 1 sidereal day. Use your observations to answer the following questions.

Question 8. What is the date of closest approach of the comet to the Sun?

Question 9. How far is the comet from the Sun at closest approach?

Question 10. What is the date of closest approach of the comet to the Earth (that is, to the observer)?

Question 11. How far away is the comet from Earth at its closest approach?

C. Further Observations of Comet Halley

The next sequences allow you to observe Comet Halley from novel perspectives.

27. Select **Favourites > Observing Projects > Comets > North Ecliptic Pole**.

In this view, you are hovering at a fixed point in space 2 AU above the Sun's surface in the direction of the north ecliptic pole on December 20, 1985, at 12:00:00 UT. The Sun and inner planets of the solar system are labeled, as is Comet Halley near the top of the view. The view also displays the orbital paths of Earth and the comet. Notice the short line perpendicular to the orbital path of the comet to the right of the Sun. This marks the perihelion point of the orbit of the comet. As the simulation begins, Comet Halley lies beyond the Earth's orbital distance from the Sun. From this fixed viewpoint in space, the objects in the solar system move, while the background stars remain stationary, as time advances.

28. With the **Time Flow Rate** in the toolbar set to **6 hours, Run time forward** and watch the comet progress through the inner planetary system. To see the animation again, select **Edit > Undo Time Flow** and then click **Run time forward.** You may also want to **Zoom In** so that Earth's orbit just fits in the view and repeat the animation once more.

Running time forward with the time interval of 6 hours will show the comet's tail grow and shrink again under the varying influence of the Sun's heat as the comet traverses the inner solar system.

Question 12. Is Comet Halley orbiting the Sun in the same direction as the planets or in the opposite direction?

Question 13. Near what point of the comet's orbit does its tail appear longest?

Question 14. On what date does the comet appear to cross the Earth's orbit on its way to the inner solar system?

Question 15. On what date does the comet appear to cross the Earth's orbit as it leaves the inner solar system?

Question 16. In which part of the comet's orbit does it come closest to the Earth, incoming or outgoing?

Question 17. a) To which planet does the comet appear to pass closest during this orbit?

b) On which date does Comet Halley come closest to this planet?

c) How close does the comet get to this planet? (Hint: The Angular Measurement Tool displays the physical distance between two objects in the view as well as their angular separation).

d) How many days does it take for Comet Halley to reach its perihelion point from this time?

e) What do you think Comet Halley would look like in the sky of this planet on this date compared to its appearance from Earth at the time of the comet's closest approach to our planet? In your answer, explain what factor or factors led you to your conclusion.

29. [Optional] To check your answer to the last question, set the **Date** in the toolbar to the date at which **Comet Halley** makes its closest approach to the planet that you identified in the last question. Position the cursor over the planet, bring up the **Object Contextual Menu** and select **Go There.** Use the **Hand Tool** to look around the sky until you find the comet. **Select File > Revert** and set the **Date** in the toolbar to the date of the closest approach of the comet to Earth. Then move the cursor over the image of the **Earth** in the view, bring up the **Object Contextual Menu** and select **Go There.** Use the **Hand Tool** to look around the view until you find the comet in the sky. [Note that the **Go There** command puts you at an elevation of about 25,000 kilometers above the surface of the planet, but the comet's appearance will not be significantly different from what it would be from the surface].

In the next sequence, you can use the location scroller to view the passage of Comet Halley through the inner solar system from different perspectives.

30. Select **File > Revert**.

31. Position the **Hand Tool** near the bottom center of the view and activate the **Location Scroller** by holding down the **Shift** key and then the mouse button. Drag the **Location Scroller** straight upward. Do this several times until the Earth's orbital path is seen edge-on horizontally across the view, and bisecting the Sun.

32. **Run time forward.** Select **Edit > Undo Time Flow** and use the time flow controls in the toolbar to manipulate time back and forth to review parts of the animation.

Question 18. On what date does the comet pass southward through the ecliptic plane?

33. [Optional] Select **File > Revert** and use the **Location Scroller** to look at the view from an oblique perspective and watch the animation as you **Run time forward.**

34. [Optional] Select **File > Revert** and use the **Object Contextual Menu** for **Comet Halley** to **Centre** this object in the view. Then use the **Zoom** and **Time Flow** controls in the toolbar to observe the comet from different perspectives.

D. Comet Orbits

In this section, you will explore some of the orbital characteristics of a sample of comets.

35. Select **Favourites > Observing Projects > Comets > Orbits**.

The view is centered on the Sun from a fixed point in space about 71 AU from the Sun. The orbital paths of the planets of the solar system and the paths of the three comets you observed in Part A above are shown in the view.

36. Use the **Location Scroller** to look at the orbital paths of these comets from different perspectives.

37. *Starry Night Enthusiast*™ can display many other comet orbits. Open the **Find** pane and expand the **Comets** layer. Click the checkbox to the right of the name of any of the long list of recent comets in the list to display its orbital path around the Sun.

Question 19. From your observations, how do the orbital paths of comets differ from those of the planets?

E. Kepler's Roller Coaster: A Ride on Comet Halley

In this section, you will use the orbit of Comet Halley to explore Kepler's second and third laws of planetary motion.

38. Select **Favourites > Observing Projects > Comets > Riding Halley's.**

39. Type **Ctrl-L** to open the **Viewing Location** dialog box. Notice that your location is on the surface of Comet Halley. Click the **Cancel** button to remain at this location.

You are viewing the solar system from the surface of Comet Halley with the gaze centered on the Sun and the horizon removed from the view. It is September 1947. The current distance of the comet from the Sun is displayed in the upper right corner of the view.

40. Notice that the **Time Flow Rate** in the toolbar is set to **10 days.**

41. **Run time forward** and observe your motion and distance from the Sun (as shown in the upper right corner of the view) as you approach, pass through, and recede from the inner solar system, riding Comet Halley in its orbit around the Sun.

42. Select **File > Revert** and choose a star near to the Sun, such as Theta Aquilae, the first bright star to the upper right of the Sun in the view. Use the **Object Contextual Menu** to **Centre** onto this star. **Run time forward** to experience the slingshot-like effect produced by the high eccentricity of this comet's orbit.

Question 20. a) Does Comet Halley move at a steady rate throughout its orbit?

b) Does Comet Halley move more rapidly when it is near aphelion or perihelion?

c) Which of Kepler's laws describes this motion?

43. Select **File > Revert** and use the **Step time forward** and **Backward** buttons to find the point at which Comet Halley is at aphelion, when its distance from the Sun, as shown in the upper right corner of the view, reaches a maximum. Record this distance in Data Table 1, below.

44. Click the numeral value in the **Time Flow Rate** panel of the toolbar to highlight it. Then click **Run time forward.** When you start to get near to the Sun, slow the **Time Flow Rate** by pressing the − key on the keyboard.

45. **Stop Time** when you notice that the **Distance to the Sun,** displayed in the upper right corner of the view, starts to increase, indicating that you have gone beyond perihelion, the closest point in the orbit to the Sun.

46. Change the **Time Flow Rate** in the toolbar to **6 hours** and **Step time backward** to find the perihelion point of the orbit, when the **Distance to the Sun** is a minimum. Record this distance in Data Table 1 below.

The sum of the two distances you measured in the previous sequence is equal to the length of the long axis of Comet Halley's elliptical orbit. One-half of this sum is equal to the semi-major axis of this orbit. With this information, we can test Kepler's third law of planetary motion, which relates the length of the semimajor axis, a, to the period, P, of the orbit by the equation

$$a^3 = P^2$$

(Equation 1)

where a is expressed in AU and P in years. Taking the square root of both sides of Equation 1 gives

$$P = a^{3/2}$$ (Equation 2)

You can use Equation 2 to determine the period that Kepler's third law predicts for Comet Halley from your measurement of its semimajor axis and then test Kepler's third law by using this period to predict the next dates at which Comet Halley will return to perihelion.

Data Table 1. Length of Semimajor Axis and Period of the Orbit of Comet Halley

Distance to Sun at Aphelion (AU)	
Distance to Sun at Perihelion (AU)	
Sum of Distances (AU)	
Length of Semi-major Axis, a (AU)	
Predicted Period, P (Years)	
Predicted Period (Years and Days)	
Year of Current Perihelion	
+ Number of Years in Period	
= Year of Next Perihelion	
Month and Day of Current Perihelion	
+ Extra Number of Days in Period	
= Month and Day of Next Perihelion	

47. Use Equation 2 to find the period, in years, of Comet Halley that is predicted by Kepler's third law. Ignoring the decimal fraction of the result for the moment, add the number of years to the **Year** of the **Date** currently in the toolbar.

48. Type the value of the **Year** for the predicted return of Comet Halley to perihelion into the appropriate field of the **Date** display in the toolbar.

49. Multiply the fractional number of years that you calculated for the period of Comet Halley by 365.25 to find the number of days that you will need to advance time to get to the perihelion point of the orbit, as predicted by Kepler's third law. Use the **Step time forward** and **Backward** buttons with a **Time Flow Rate** of **1 day** to test whether the prediction of the next date of perihelion for Comet Halley was correct.

Question 21. What is the period of Comet Halley in years?

Question 22. What date did Kepler's Law predict for the next perihelion of Comet Halley?

Question 23. When did the next perihelion actually occur?

Question 24. Do your results verify Kepler's third law?

F. The Speed of Comet Halley

In this section, you can measure the speed of Comet Halley at two points in its orbit. You can make these observations from the center of the Sun. The two points at which you will measure the speed of the comet are when it is near perihelion and near aphelion. Thus, you will measure the comet's speed when it is moving fastest and slowest, respectively. Another important reason for choosing these points in the comet's orbit is that it is moving at right angles to the line of sight at these points. Thus, you can measure its velocity by observing its angular change in position in the sky without needing to account for projection effects. You can thus verify Kepler's second law of planetary motion by using your observations and measurements to calculate the area swept out by the comet in a specific period of time at these points in its orbit and then compare these two areas.

50. Select **Favourites > Observing Projects > Comets > Speed at Perihelion.**

The view shows Comet Halley and its orbital path as seen from the center of the Sun in February 1986 when the comet is at the perihelion point of its orbit.

51. **Zoom In** to the minimum field of view of < 1'' wide.

The mark on the orbital path of Comet Halley just to the right of the comet in the view is the perihelion point of its orbit.

52. Measure the angular separation between the center of the comet and the perihelion mark on its orbital path and make a note of this separation.
53. **Step time forward** by 1 step of **2 seconds**. During these 2 seconds, the comet passes perihelion, but since the gaze is centered upon and follows the comet as it moves across the sky, the perihelion point appears on the other side of the comet on the screen. Measure the angular separation between the center of the comet and the perihelion mark at this time. Add this measurement to the previous separation and record the result in Data Table 2 below under Angular Motion of Comet in Interval, α.
54. Position the **Hand Tool** over the image of the comet and record the **Distance from Observer** as displayed in the **HUD** into Data Table 2 for this date. Since you, the observer, are at the center of the Sun, the **Distance from Observer** is equal to the comet's distance from the Sun at this point in its orbit.
55. Select **File > Revert** and then change the **Date** in the toolbar to **April 24, 2024** AD.

About thirty-eight years after reaching perihelion, Comet Halley has moved to the outer reaches of the solar system and is near aphelion, its farthest distance from the Sun.

56. **Zoom In** to a field of view about 2' wide and notice the star to the right of the comet in the view. Use the **Hand Tool** to identify this star as **TYC199-1178-1** in the **HUD**.
57. Open the object contextual menu for this star and select **Centre.**
58. Change the **Time Flow Rate** in the toolbar to **10 days.**
59. **Step time forward** once and note the direction in which Comet Halley moves in this interval of time. (The orbital path of Comet Halley is very dim in this display but it is just visible if you look closely along the direction in which the comet moved.)
60. **Step time backward** once so that the date is again **April 24, 2024** AD.

61. Select **View > Celestial Grid** to bring up a reference grid attached to the celestial sphere. Since the gaze in the view is centered on a star, this reference grid will remain essentially stationary in the view when you advance time in the next step. Select the first vertical grid line to the left of the comet and measure the angular separation between the comet and this reference line in a direction along the line of Comet Halley's movement (parallel to the orbital track dimly displayed in the view). Make a note of this measurement.

62. **Step time forward** by one interval of **10 days** and measure the angular distance from Comet Halley to the same reference gridline you used in the previous step. Find the difference between these two measurements in arcseconds and record the result as the motion of Comet Halley in arcseconds at the aphelion point of its orbit over an interval of 10 days in Data Table 2.

63. Use the **Hand Tool** and **HUD** to find the **Distance from Observer** of Comet Halley on this date and record this distance in Data Table 2.

Data Table 2. Speed and Distance from the Sun of Comet Halley at Perihelion and at Aphelion

Position of Comet Halley in its Orbit	Time Interval (s)	Angular Motion of Comet in Interval, α (")	Comet's Distance from Sun, d (AU)	Distance Comet Moved in Interval (km)	Angular Speed of Comet, ω ("/s)	Product of $d^2 \times \omega$
Perihelion	2					
Aphelion	864000					

Before using your observations and data to verify Kepler's second law, it is interesting to consider the speeds that these angular movements represent in more familiar terms such as kilometers per second. To do this, you need to use the small angle formula, $D = \alpha \times \dfrac{d}{206265}$, to convert the angular motion of the comet in the interval, α, into distance in kilometers. The distance of the comet, d, is in AU, so D will be in AU in this formula. Since 1 AU = 1.496×10^8 km, the value of D in km is given by the equation

$$D = 725.3 \times \alpha \times d \qquad \text{(Equation 3)}$$

where D is the distance in kilometers moved by the comet during the interval, α is the angular motion of the comet in the interval, in arcseconds (column 3 of Data Table 2) and d is the distance of the comet from the Sun in AU (column 4 of Data Table 2).

64. For the perihelion and aphelion points of the comet's orbit, use Equation 3 to determine the distance the comet moved during the interval, in kilometers, and record this in Data Table 2 above (column 5). Now you can find the speed of the comet at these two points in its orbit in units of kilometers per second by dividing this distance by the number of seconds in the interval over which the motion was measured (column 2 of the data table).

Question 25. What is the speed, in kilometers per second, of Comet Halley at (a) perihelion, and (b) aphelion?

Question 26. How much faster does Comet Halley move at perihelion compared to its speed at aphelion?

Question 27. How much faster does Comet Halley move when it is at perihelion compared to the average speed of the Earth in its orbit (29.79 km/s)?

[In terms of everyday speeds, we can translate these values from km/s to km/h by multiplying by the number of seconds in an hour, namely 3600. To translate these speeds to mph, divide the speeds in km/h by 1.6. You can see that, on its close approach to the Sun, this celestial visitor is traveling very fast indeed and could inflict great damage on any planet in its path!]

As you have seen, Comet Halley moves through a wide range of distances from the Sun as it traverses its orbit. Thus, we can use the measurements of its motion that you recorded in Data Table 2 above to verify Kepler's second law over this range of distances. Kepler defined this law in the following terms: "A line joining a planet and the Sun sweeps out equal areas in equal intervals of time."

As you recall, the interval you used to measure the angular motion of Comet Halley when it was at aphelion was much longer than the interval that you used to measure the angular motion of this comet at perihelion. Since Kepler's second law stipulates equal intervals of time, you must convert your measurements of the comet's angular motion to a common interval of time.

65. Divide the measured angular motion of the comet, α, (column 3 of Data Table 2) by the number of seconds in the interval (column 2 of Data Table 2) to express the angular motion of the comet at both points of its orbit in common units of arcseconds per second of time and record this as the angular speed of the comet, ω, in column 6 of Data Table 2.

The relationship between the area swept out by the comet and measurable quantities can be derived easily. The relevant observations are the angular speed of the comet against the background stars, and the distance of the comet from the Sun, both of which you have obtained in Data Table 2.

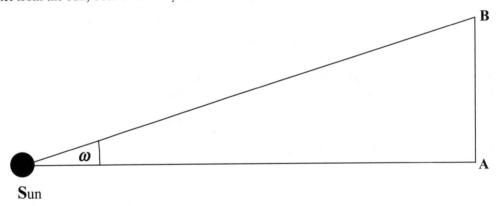

Figure 1. Simplified Geometry of Area of a Section of an Orbit

Since the angular distances that you measured for the motion of Comet Halley were very small, and the motion of the comet at the times of observation near perihelion and aphelion was across the line of sight, we can approximate the motion of the comet with a right triangle as illustrated in Figure 1 above.

At the perihelion and aphelion points of its orbit, the angular distance that the comet moves in one second, say from B to A in the figure, is equal to ω, its angular speed in arcseconds per second at that point in its orbit.

The area of triangle SAB is equal to one-half of the product of its base (AB, the angle through which the comet appears to move across the sky in one second), and its height (SA, the distance from the Sun to the comet). If we let $d = SA$, then

$$AB = d \tan(\omega)$$

When ω is small, as it is in the present case, then $\tan(\omega) \approx \omega$ if ω is expressed in radians. Then

$$AB \approx d \times \omega$$

So, the area of triangle SAB

$$= \frac{1}{2} \times SA \times AB$$

$$= \frac{1}{2} \times d \times d \times \omega$$

$$= \frac{1}{2} \times d^2 \times \omega$$

where ω is expressed in radians.

With the distance to the comet, d, expressed in AU, the area in units of AU^2 swept out by the comet in one day is

$$= 1/2 \times d^2 \times \omega \text{ (arcseconds)} \times (1 \text{ radian} / 206,265 \text{ arcseconds})$$

$$= 2.42 \times 10^{-6} \times d^2 \times \omega$$

$$= \text{constant} \times d^2 \times \omega$$

Thus, to verify Kepler's second law that the area swept out by the sun-comet line per unit time is constant, we need to show that the product $d^2 \times \omega$ remains constant for Comet Halley along its orbit.

66. To see whether the line from the Sun to Comet Halley sweeps out equal areas at the two extreme points in its orbit and so obeys Kepler's second law, calculate the product of the square of the distance of the comet from the Sun, d (which you recorded in column 4 of the data table) and the angular speed of the comet at that point in its orbit, ω, which you calculated in the previous step and entered into column 6 of the data table. Enter this product into column 7 of Data Table 2.

Question 28. Do the two values that represent the area swept out by Comet Halley at the two points of its orbit agree? What do you conclude from this?

G. Conclusions

You have explored the motion and changing appearance of Comet Halley on its last apparition in our sky in 1985 and 1986. You have watched this motion from several locations, some not accessible in real life. You made observations from these unique locations that allowed you to verify that the comet obeys Kepler's laws of planetary motion. From these measurements, you have also been able to calculate values for the tremendous speed that this comet acquires as it sweeps through our solar system, and you have demonstrated that it moves much faster when it is near perihelion than at any other time in its orbit.

Parallax

17

arallax is the apparent shift in the position of a relatively nearby object against a more distant background that is caused by a change in an observer's position. This effect is the foundation on which our knowledge of the size of the universe, and the distances between objects and structures

within it, is built. The measurement of distance to objects in the universe is crucial to the determination of many other parameters of importance in astrophysics. For example, the luminosity of a star, its total energy output, can only be deter-

Parallax is discussed in Section 17–1 of Freedman, Geller, and Kaufmann, *Universe*, 9th Ed.

mined from the measured intensity of the light that reaches us if we know its distance from the Earth.

A simple experiment demonstrates the parallax effect. Close your right eye and note the position of your outstretched hand against a distant background when observed along the line of sight of your left eye. Now close your left eye and open the right. The difference in the line of sight between your two eyes causes your hand to appear to "jump" to the left against the background. The closer an object, the more pronounced the effect. You can verify this by repeating the experiment with your hand held closer to your face.

The parallax effect also depends on the distance between the two viewing points; that is, the length of the observational baseline. In the experiment above, the baseline is the distance between your two eyes (specifically, the pupil-to-pupil distance). If you try to repeat the experiment by observing an object much farther away, for example, a tree trunk or signpost about 30 meters away, and compare its position against an even more distant background when viewed through each eye independently, it is unlikely that you will notice any parallax. However, if you note the position of the tree trunk or signpost against the background from one spot and then move several meters to the right or left, the parallax effect becomes obvious once more.

The form of parallax described above is called **trigonometric parallax.** When it is applied to astronomical objects, the distance to the object of interest is generally much greater than the observational baseline. The angle, θ, in radians, through which an object at a distance d shifts against a much more distant background, when observed from two different locations a known distance, X, apart and measured in the same units as d, is then given by the small-angle formula:

$$\theta = X / d$$

(Equation 1)

As you can see from Equation 1, the parallax angle of an object, θ, is directly proportional to the length of the observational baseline, X, and inversely proportional to the distance to the object, d. Two interesting and very useful consequences of this result are:

a. For an object at a given distance, increasing the length of the observational baseline produces a larger parallax angle, thus increasing the accuracy of the measurement of the distance to this object.

b. For a given observational baseline, the more distant an object, the smaller is its parallax angle; thus, for a given baseline (e.g., the diameter of the Earth), the minimum measurable angle of parallax determines the maximum measurable distance to an object.

187

Parallax angles in astronomy are usually very small and are expressed in arcseconds, whereas the formula above assumes that the angle is expressed in radians. For convenience, therefore, astronomers usually rewrite Equation 1 with a factor to convert radians to arcseconds. An angle of 2π radians is a full circle, or 360°, and a full circle is $360 \times 60 \times 60$ arcseconds. Thus,

1 radian = $360 \times 60 \times 60$ arcseconds / 2π radians = 206,265 arcseconds

If we measure the parallax angle of an object in arcseconds, we can rearrange Equation 1 to determine the distance to the object as

$$d = 206{,}265\ X\ /\ \theta$$

(Equation 2)

where θ must be in arcseconds, and the distance, d, will be in the same units as the observational baseline, X.

A. Parallax of the Moon

In this section, you can observe and measure the parallax shift of the Moon by using two observing sites on opposite sides of the Earth. The Moon will be setting at one location and just rising from the second location.

1. Launch *Starry Night Enthusiast* ™.

2. If a side pane is open, close it to maximize the view.

3. Select **Favourites > Observing Projects > Parallax > Moon from 90W**.

4. Select **File > New**.

5. In this new *Starry Night Enthusiast*™ application window, use the menu to select **Favourites > Observing projects > Parallax > Moon from 90E**.

The date is November 9, 2004, and these two views are from locations on the equator of Earth at longitudes of 90° W and 90° E, respectively. The local times are 3:08:06 AM and 3:08:06 PM at these observing sites. The date and time have been specifically chosen so that the Moon is near to the celestial equator and is transiting the meridian as seen from the prime meridian on Earth at longitude 0°, half-way between the two observing positions. The Universal Date and Time at this longitude are displayed in the top right corner of the view. The gaze is fixed on a point on the celestial sphere (marked by a dim star, TYC282-822-1, which is at the center of the view but not labeled). The field of view is 5° wide by 3° high with north at the top and east to the left. Within this field, a waning crescent Moon is visible against the background of the celestial sphere.

6. Flip back and forth several times between the two views by alternately selecting **Moon from 90E** and **Moon from 90W** under the **Window** menu.

You can see that the Moon appears to move through a significant angle in the sky when viewed from these diametrically opposed locations on the Earth.

Question 1. When you flip the view from the location at longitude 90° E to the view from the location at 90° W, in which compass direction (east or west) does the Moon appear to move?

Question 2. What is the time difference between the two observing locations?

Question 3. As seen from longitude 90° E, is the Moon rising or setting? [*Note:* The horizon is hidden in these views but can be displayed by clicking the checkbox next to **Local Horizon...** under the **Local View** layer of the **Options** pane.]

Question 4. For an observer on the equator of the Earth at the prime meridian, a) what would be the local time? b) At which specific point in the sky would the Moon be for this observer?

7. Select **Window > Moon from 90E** from the menu.

8. If a side pane is open, close it.

9. Select **File > Revert**.

10. Select **Window > Moon from 90W** from the menu.

11. If a side pane is open, close it.

12. Select **File > Revert**.

One specific star, TYC281-1009-1, can be seen to the right of the Moon from both viewing locations. This star is labeled and will serve as our reference star for measuring the parallax shift of the Moon.

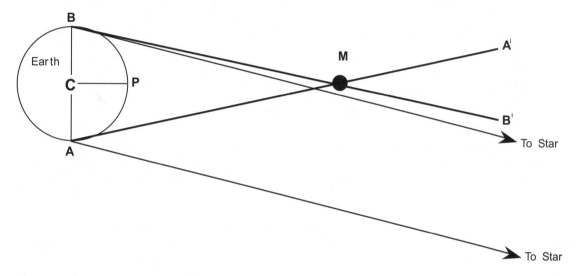

Figure 1. Parallax Geometry

Figure 1 illustrates the geometry of the parallax effect that you will measure and use to calculate the distance to the Moon. The size of the angle AMB in the diagram is exaggerated. In practice, this angle is very small. This diagram shows a cross-section through the center of the Earth in an equatorial plane that contains the Moon. The Moon, at point M, is on the meridian for an observer at point P, at longitude 0° on the Earth. The Earth is rotating counter-clockwise from this view. The two observing points are at A, at longitude 90° W, and at B, at longitude 90° E, respectively. The reference star is assumed to be at infinity compared to the finite distance to the Moon and can be considered to be on the celestial sphere. The two lines of sight from A and B to the star will be parallel, as shown in the diagram. The reference star has been chosen just to the right, or west, of the projected positions A' and B' of the Moon on the celestial sphere. Thus, you can measure the angular separation between the apparent position of the Moon at A' and the reference star as seen from position A and then move to the second viewing location at B and measure the angular separation between the apparent position of the Moon at B' and the reference star. The difference between these two angles is the parallax angle of the Moon when measured over the baseline of the Earth's diameter.

This angle is equivalent to the angle subtended by the diameter of the Earth when seen from the Moon. This equivalence is easier to visualize if you imagine the reference star to be coincident with the center of the Moon when viewed from position B. In this ideal case, the angle between the star and the Moon's apparent position when seen from A will be the parallax angle. The sight-lines to the star from A and B are parallel and line AM crosses these parallel lines. In this situation, opposite angles are equal. Thus, in this ideal situation, the angle subtended by the Earth's diameter, angle AMB, is equal to the angle between star and Moon when seen from A. The small additional offset that places the reference star at the side of the Moon is necessary to allow the measurements of the Moon's positions from these two observing sites to be made, and is removed by subtraction in the following method.

13. In the **Moon from 90W** window, use the **Angular Measurement Tool** to measure the angular separation between the center of the Moon and the labeled reference star TYC281-1009-1. Record this measurement in Data Table 1 below.

14. Select **Window > Moon from 90E** and use the **Angular Measurement Tool** to measure the angular separation between the center of the Moon and the labeled reference star TYC281-1009-1 from

Data Table 1. Parallax Shift of the Moon Against the Background Stars

Longitude of Viewing Location	Angular Distance between Moon and TYC280-1009-1		
	°	'	"
90° West			
90° East			
Parallax Shift			

15. In Data Table 1, subtract the measurement of the angular separation between the center of the Moon and the reference star as observed from longitude 90°E from the measurement obtained from the viewing location at longitude 90°W and record the difference in the row labeled **Parallax Shift.**

The difference between these two measurements is the parallax shift of the Moon produced by moving from 90°W to 90°E on the Earth at this time on this date. This angular measurement corresponds to θ in Equation 2 and is equivalent to the angle AMB in Figure 1.

The observational baseline between points A and B in Figure 1 that produced this parallax shift is the distance between the viewing locations, X in Equation 2, and is equal to the diameter of the Earth, 12,756 km. Finally, the distance from the center of the Earth to the center of the Moon, which corresponds to d in Equation 2, is the distance between point C and point M in Figure 1.

Question 5. What angle does the observational baseline (AB in Figure 1) make with the line from the center of the Earth to the center of the Moon (CM in Figure 1)?

16. Convert the parallax shift of the Moon that you measured in Data Table 1 to arcseconds.

Question 6. What is the parallax shift of the Moon on this date, produced by observing it from these two opposite locations, in arcseconds?

Question 7. Using Equation 2 and the parallax shift you measured, what is the distance to the Moon, in kilometers?

17. Open the object contextual menu for the **Moon** and select **Show Info.** Look in the **Position in Space** layer of the **Info** pane and compare the value given for **Distance from observer** with your calculation of the distance to the Moon from its parallax shift.

Question 8. How does your calculation for the distance to the Moon compare to that given in the **Info** pane of *Starry Night Enthusiast*™?

B. Geocentric Parallax

Parallax from a baseline limited to the diameter of the Earth is called **geocentric parallax**. The strict definition of geocentric parallax is the angular shift in position of an object when observed from a location on the surface of the Earth in a tangential direction to the Earth's surface when compared to the equivalent, hypothetical observation made from the center of the Earth. The baseline would then be the radius of the Earth. This is equivalent to the parallax shift that would be seen by making an observation of the position of the Moon against the celestial sphere from either location A or B in Figure 1, compared with an observation from a point at the center of the Earth, point C in the figure. As you can see from the figure, this would be exactly one-half of the parallax shift that you measured along the baseline of the diameter of the Earth.

> **Question 9.** What is the geocentric parallax, in arcseconds, of the Moon on the date observed in the previous sequences?

Stated another way, the geocentric parallax of an object is equal to the angular radius of the Earth as seen from the distance of that object. You can demonstrate this equivalence with the following simulation.

18. Select **Window > Moon from 90W**.

19. Open the object contextual menu over the center of the image of the **Moon** and select **Go There**. Press the **Spacebar** to go instantly to a viewing location at 6951 km above the Moon's surface.

20. Open the **Find** pane and click the menu button next to the listing for **Earth** and select **Magnify** from the dropdown menu.

21. Use the **Decrease current elevation** button in the toolbar to lower the observing location to the surface of the Moon.

The view shows the Earth as seen from the Moon on the date and at the time that you measured the parallax shift of the Moon from the Earth in the previous sequences.

22. Measure the angular radius of the Earth in this view.

> **Question 10.** a) What is the measured angular radius, in arcseconds, of the Earth as seen from the Moon at this point in time?
>
> b) Within the precision of your ability to measure the angular radius of the Earth as seen from the Moon, how well does your measurement of the radius of the Earth agree with your answer to the previous question?

> **Question 11.** Assume that you are using equipment for measuring parallax that can detect angular displacements as small as one-half of an arcsecond (0.5'') and that the observational baseline you will use is equal to the diameter of the Earth (12,756 km).
>
> a) What is the distance, in kilometers, to the farthest object for which you could measure a parallax shift?
>
> b) What is this distance expressed in AU?
>
> c) What is the farthest planet in the solar system for which you could measure parallax, using this Earth baseline?

C. Geocentric Parallax of Vesta

Distances in the solar system beyond that of the Moon are more usefully expressed in astronomical units (AU) than in kilometers. To find the distance to an object in the solar system in AU by measuring its parallax shift in arcseconds, you will also need to express the baseline distance, X in Equation 2, in terms of AU.

This means that you will have to convert the diameter of the Earth, the baseline you will use for measuring the parallax of Vesta, from kilometers to AU. Since 1 AU = 1.496×10^8 km and the diameter of the Earth is 12,756 km, this conversion factor is:

$$12{,}756 / 1.496 \times 10^8 = 8.527 \times 10^{-5}$$

Combining this with the factor 206,265 for converting radians to arcseconds into a single constant simplifies Equation 2 to:

$$d \text{ (AU)} = 17.59 / \theta \text{ (arcseconds)} \qquad \text{(Equation 3)}$$

You can use the method of trigonometric parallax to measure the distance to the asteroid, Vesta. *Starry Night Enthusiast*™ provides a convenient, reliable and accurate method for measuring this parallax by displaying a Field of View circle of appropriate size on the sky surrounding Vesta from one location. This circle will remain in the same position when the observing location is changed. The change in Vesta's apparent position can be measured accurately by referring to this circle of known radius.

First, you can examine Vesta's position from two locations on the Earth 180° apart at the same Universal Time.

23. Select **Window > Moon from 90E**. If a side pane is open, close it and then use the menu to select **Favourites > Observing Projects > Parallax > Vesta from 59E.**

24. Select **Window > Moon from 90W**. If a side pane is open, close it and then use the menu to select **Favourites > Observing Projects > Parallax > Vesta from 121W.**

The current view is from the equator of Earth at longitude 121°W, showing the asteroid Vesta near the top, or north edge, of the view. The view is 5 arcminutes wide and is centered and locked onto the star TYC291-188-1. The local time at this location is 4:29:17 AM on October 28, 2006. This time and date have been chosen so that Vesta is transiting the meridian at a longitude directly between the longitudes of the two views. The equatorial reference grid is displayed against the background sky.

25. Select **Window > Vesta from 59E.**

At this location, separated from the previous location by 180° of longitude, the local time is 4:29:17 PM on October 28, 2006.

Question 12. What is the distance between the two viewing locations?

Question 13. At what longitude would you need to be to see Vesta transit the meridian at this time on this date? [*Hint:* Look at Figure 1.]

Question 14. What is the length, in AU, of the observational baseline established by these two viewing locations?

Question 15. a) How would the observational baseline, established by these two viewing locations, change if they were both at latitude 60° north?

b) What would be the length of the observational baseline in AU if the viewing locations were both at latitude 60° north?

Question 16. Where is Vesta relative to the celestial equator in these views? [*Hint:* Look at the coordinate grid in the view.]

26. Flip back and forth between the views to observe the parallax shift of Vesta by using the menu to alternately select **Window > Vesta from 121W** and **Window > Vesta from 59E.**

Since the gaze in both views is centered on the star TYC291-188-1, it is locked onto a specific point of the celestial sphere. The celestial grid, which is attached to the celestial sphere, and the star remain unaffected as you change the viewing location from one end of the observational baseline to the other. Vesta, however, shows a definite parallax shift.

You can now measure this shift between the views from the two different locations, using a convenient Field of View circle centered on the position of Vesta when viewed from one location.

27. Select **Window > Vesta from 59E.** Use the object contextual menu to **Centre** the view on **Vesta** in this window.

28. **Zoom In** to a field of view about **12''** wide.

29. Open the object contextual menu for Vesta and select **Add FOV Indicator > Circular...**

30. In the dialog window that appears, Check the **Positioning** box and ensure that **RA/Dec** is selected. This will keep the FOV circle fixed on the sky when the apparent position of Vesta changes.

31. In the **Diameter** box, enter 0.003 degrees. This angle is equal to $0.003 \times 60 \times 60$ arcseconds = 10.8'', so the radius of this circular FOV is 5.4''. Click **OK** to display this circle around Vesta. You can use the **Hand Tool** to check that the radius of this circle is 5.4''.

You now need to change observing location to move to the opposite side of the Earth by adding 180° to the longitude of the present observing location.

32. In the **Options** menu, select **Viewing Location...**

33. In the **Viewing Location** dialog window, click the **Latitude/Longitude** tab. The present longitude value is 58° 42.54' E. Adding 180° to this value leads to a new longitude of 238° 42.54' E. Change **Longitude** to this value and click the **Set Location** button. Press the **spacebar** to change locations instantly.

34. Open the **Find** pane and type **Vesta** in the search box followed by the **Enter** key. Again, pressing the **spacebar** will move the view quickly to center upon Vesta. If the FOV circle does not appear, you may need to **Zoom Out** to a somewhat larger field of view.

You will see that the FOV circle that was centered on Vesta's position from the first location is now offset from Vesta. The radius of this circle is 5.4''. Thus, you can measure the parallax movement of Vesta from the first location to the second by simply measuring the angle between Vesta and the nearest point on the FOV circle. The radius of the circle was chosen so that this angle would be sufficiently small that the enhanced precision of the Hand Tool in *Starry Night Enthusiast*™ would come into operation, providing measurements with a precision as high as 0.01''.

35. Use the **Hand Tool** to measure the angular spacing between the center of Vesta and the nearest point on the FOV circle and add this value to 5.4'' to calculate the change in position of Vesta due to parallax when location changes by 180° in longitude on the Earth.

36. Use Equation 3 to calculate the distance from the Earth to Vesta in AU.

37. Open the contextual menu for **Vesta** and select **Show Info.** Find the value that *Starry Night Enthusiast*™ gives for the **Distance from observer** of Vesta under the **Position in Space** layer of the **Info** pane.

Question 17. a) What is the parallax shift of Vesta from the above measurements? b) What is the geocentric parallax of Vesta?

Question 18. What distance, in AU, did you calculate for Vesta based on your measurements of its parallax?

Question 19. How does your result compare with the value for the distance to Vesta given in *Starry Night Enthusiast*™?

D. Parallax of Other Solar System Objects

It is possible to apply the same method of parallax measurement to other objects in the solar system. For each of the objects listed in Data Table 2, follow the instruction sequence below to measure the parallax shift of the object and determine its distance.

Data Table 2. Parallax and Distance from Earth of Several Solar System Bodies

View	Suggested FOV Circle Diameter	FOV Radius, R	Measured Angle to Circle, Δ (")	Parallax Shift, $R + \Delta$ (")	Calculated Distance (AU)	Actual Distance, From Info Pane (AU)
Europa from 96W	5"	2.5"				
Iapetus from 126W	3"	1.5"				
Miranda from 7W	1"	0.5"				
Nereid From 44W	0.5"	0.25"				

38. Use **Favourites > Observing Projects > Parallax** and select the specific View indicated in Data Table 2 for the chosen object.

39. **Zoom In** to a field of view of about 12 arcseconds.

40. Open the object contextual menu and select **Add FOV Indicator > Circular...**

41. Check the **Positioning** box and ensure that **RA/Dec** is selected, in order that the FOV circle remains fixed on the sky.

42. In the **Diameter** box, enter the value of the **Suggested FOV Circle Diameter** appropriate to the chosen object, as listed in Data Table 2. (The "double-quotes" key on the keyboard should be used to indicate arcseconds.) The relevant value for the radius R of this reference circle in arcseconds is shown in the data table. Click **OK** to display this circle around the object. You can use the **Hand Tool** to verify this radius value.

You can now change your observing location to the opposite side of the Earth by adding 180° to the longitude of the present observing location.

43. Select **Options > Viewing Location...** and click the **Latitude/Longitude** tab. Add 180° to the longitude value shown in the **Longitude** box and replace the "degrees" number with this new value. Click on **Set Location** and press the **spacebar** to move to this new location quickly.

44. Open the **Find** pane; type the name of the chosen object in the search box followed by the **Enter** key. (If the FOV circle, offset from the object, does not appear, you may need to **Zoom Out** to a somewhat larger field of view.)

You will see that the FOV circle that was centered on the object's position from the first location is now offset from the object. You can now measure the parallax movement of the object from its initial position to its present position by simply measuring the angle between the object and the nearest point on the FOV circle, labeled Δ in the Data Table, and adding the radius R of this circle. As in the case of Vesta, the radius of the circle has been chosen so that this angle would be small enough that the **Hand Tool** in *Starry Night Enthusiast*™ can measure this residual angle to high precision, in fact, to 0.01".

45. Use the **Hand Tool** to measure the angular spacing between the center of the object and the nearest point on the FOV circle and enter this value in Data Table 2, in the **Measured Angle to Circle, Δ** column. Add this value to the radius R of the FOV to calculate the change in position of the object due to parallax when location changes by 180° in longitude on the Earth and enter the sum in the **Parallax Shift, R + Δ** column of Data Table 2.

46. Use Equation 3 to calculate the distance from the Earth to the object in AU and enter this value in the **Calculated Distance** column of the data table.

47. Open the object contextual menu for the chosen object and select **Show Info.** You can find the actual distance of this object under **Distance from observer** in the **Position in Space** layer of the **Info** pane. Enter this value in Data Table 2 under **Actual Distance.**

48. Repeat the sequence of steps beginning at instruction 38 for each of the views listed in Data Table 2.

Question 20. What distance did you calculate from the parallax shift of a) Europa, b) Iapetus, c) Miranda, and d) Nereid?

From these measured values and from those for the Moon and Vesta you can see that excellent accuracy can be obtained with this method when used in this simulation. In real life, very careful observations are needed in order to achieve high precision and this method of distance measurement has been superseded by other, much more precise methods utilizing space techniques, since spacecraft have visited all of these distant objects except Vesta.

The factor that limits the accuracy of distance measurement by this method is the precision with which the measurement of parallax angle can be measured. In the present measurements, the FOV circle radius is assumed to have no error while the precision of the additional incremental measurements is 0.01". Using this value as the uncertainty in measurement of parallax, you can calculate the fractional error in the parallax angle to which this precision leads and apply this fractional error to the measured distances. (In fact, *Starry Night Enthusiast*™ can provide better precision than this when very small angles are being measured, but this is not the case in the present procedure.) You can calculate the potential effect of this precision on the measurement of distance by this method and check for a possible trend in these measurements with distance.

49. From the data on the Moon, Vesta, and the objects listed in Data Table 2, collect the parallax measurements and enter them in column 2, Measured Parallax, p, of Data Table 3 and the distance measurements calculated from these parallax measurements and enter them in column 4, Calculated Distance from Parallax, d. You will have to convert your calculated distance to the Moon in kilometers to AU by dividing your distance measurement by 1.496×10^8 kilometers.

50. Calculate the fractional error in the parallax measurements caused by an error of 0.01 arcseconds, $F = 0.01/p$, and enter these values in column 3 of Data Table 3. (Since the Moon measurements were made to a precision of 1 arcsecond, you should calculate the fractional error caused by this precision for the Moon.)

51. Calculate the error in AU in the distance measurements for each object caused by the measurement precision for that object by multiplying the fractional error by the calculated distance, $F \times d$, and enter these values in column 5 of Data Table 3.

52. Examine these values for possible trends between distance and error, plotting error versus distance on a graph to evaluate any trends further, if necessary.

Data Table 3. Parallax Measurement Errors

Body	Measured Parallax p (")	Fractional Error for 0.01" precision, $F = 0.01/p$ (For Moon, $F = 1/p$)	Calculated Distance from Parallax, d (AU)	Error in measured distance, $F \times d$ (AU)
Moon				
Vesta				
Europa				
Iapetus				
Miranda				
Nereid				

Question 21. Is there a relationship between error and distance for these measurements?

Question 22. What would be the maximum measurable distance to an object by this method if you place a limit of 10% on the precision of the parallax measurement?

E. Geocentric Parallax Using Shorter Observational Baselines

While the parallax measurements of the previous sections of this project were carefully designed, the luxury of making such observations is rare. In reality, parallax shifts of nearby objects are measured from shorter observational baselines. While this is more convenient from the practical standpoint of coordinating the required observations, it is somewhat more difficult conceptually and mathematically.

In this section of the project, you will measure the parallax of the Moon while moving the observing position in a north-south direction through a range of latitudes along a constant line of longitude. This longitude has been chosen to place both the Moon and a reference star on the local meridian at the time of observation. You can see the parallax effect that you will measure directly in the following sequence.

53. Select **Favourites > Observing Projects > Parallax > Geocentric Parallax.**

The view shows the lower limb of the Moon just touching the labeled star TYC1886-1179-1. This star and the Moon are both on the meridian (the vertical line on the screen) at this time. Notice that the viewing location is 425 kilometers NW of Alesund, Norway.

> 54. Position the **Hand Tool** on the meridian midway between the lower limb of the Moon and the bottom of the view. Hold down the **Shift** key to activate the **Location Scroller.** Then click the mouse button and slowly drag the mouse up and down in the view while trying to keep the Moon and star close to the meridian as you do so. The result of this action is to change your location on the Earth. You can see this in the **Viewing Location** box above the view. Because the view is centered and locked on the specific point on the distant celestial sphere marked by the reference star, the position of the Moon appears to change as you use the **Location Scroller** to move your observing location.

Question 23. In which direction does the Moon appear to move as your location moves southward?

Figure 2, below, illustrates the geometry of the viewing locations you will use in the sequences to follow.

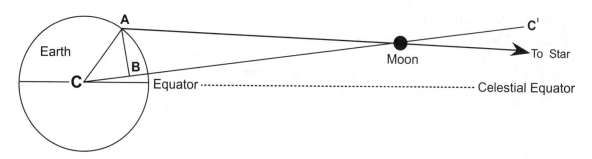

Figure 2. Geocentric Parallax Geometry

The diagram shows a slice through the Earth, coincident with the chosen longitude where the Moon and the selected star are on the meridian. The point C' shows the position on the celestial sphere where the lower edge of the Moon would appear to be to an hypothetical observer at the center of the Earth, at point C in the figure. Recall that geocentric parallax is defined as the apparent angular shift in the position of an object in the sky when observed from a location on the surface of the Earth compared to the object's position when observed from a hypothetical location at the center of the Earth. The line CC', therefore, is the reference line for determining the length of the baseline for these two observations (the "geocentric baseline"). If we choose an observing position (point A) at a given latitude along the chosen line of longitude, the geocentric baseline is the line AB perpendicular to the reference line CC'. As you can see in Figure 2, if the observing location is moved south toward the equator, this baseline will decrease in length and we would expect the geocentric parallax also to decrease. Conversely, if the observing location is moved further north, the geocentric baseline increases in length and the geocentric parallax will increase.

The strategy for measuring the geocentric parallax of the Moon will be to measure the changing angular distance between the reference star and the center of the Moon as we move progressively further south along the chosen line of longitude. You can then plot the magnitude of this angular distance at each location against the length of the geocentric baseline. In addition to measuring the angular distance between the reference star and the center of the Moon, you will also need to determine the length of the geocentric baseline for each observing location. This can be calculated from knowledge of the latitude of this location. The latitude, λ, of the observing location is the angle between a radius joining it to the Earth's center and an equivalent radius joining Earth's center to the equator on the same line of longitude. This corresponds to the angle ACE in Figure 2. The angle formed by the equator and the reference line CC' is the declination of the Moon in the sky, δ.

In triangle ABC, the angle ACB = $(\lambda - \delta)$ and the hypotenuse of this triangle, AC, is equal to the radius of the Earth, 6378 km. Thus, the length of the required baseline is

AB = AC · Sin $(\lambda - \delta)$

     = 6378 · Sin $(\lambda - \delta)$, in km.

You can now proceed to measure the change in parallax of the Moon with respect to the reference star as the latitude of the observing location is changed.

55. Select **File > Revert.**

56. **Zoom Out** to a field of view about 3° wide.

57. Open the object contextual menu for the **Moon** and select **Show Info.** Under the **Position in Space** layer of the **Info** pane, note the **Dec (JNow)** value. This is the declination of the Moon at this time on this date. Convert this to decimal degrees and record the result at the top of Data Table 4 below.

58. Open the **Status** pane and, under the **Location** layer, note the **Latitude** for this observation. Convert this value to decimal degrees and record the result in Data Table 4 using positive values for latitudes north of the equator and negative values for latitudes south of the equator.

59. Measure the angular separation between the center of the Moon and the reference star. Record the measurement in Data Table 4. Convert the measurement to arcseconds and write this value in the last column of Data Table 4.

60. Use the **Location Scroller** in a downward motion to move your viewing some distance south. Make sure that you keep the Moon and the reference star on the meridian.

61. Repeat steps 58–60 to make further observations until the Moon has moved out of the view.

62. After you have completed a number of observations over a range of latitudes from about 65° N to − 24° S, calculate the values of $(\lambda - \delta)$ and Sin$(\lambda - \delta)$ in Data Table 4.

Data Table 4. Geocentric Parallax of the Moon

Declination of Moon, δ (°)						
Latitude, λ (°)	$(\lambda - \delta)$ (°)	Sin$(\lambda - \delta)$	Angular Separation Between Moon and Star			Separation (")
			°	'	"	

63. On the graph in Figure 3 or in a spreadsheet application, plot the angular separation between the Moon and the reference star against Sin $(\lambda - \delta)$.

64. Draw the best straight line that fits through the data on this graph (or use a trendline function in a spreadsheet application such as Excel). Determine the slope of this line.

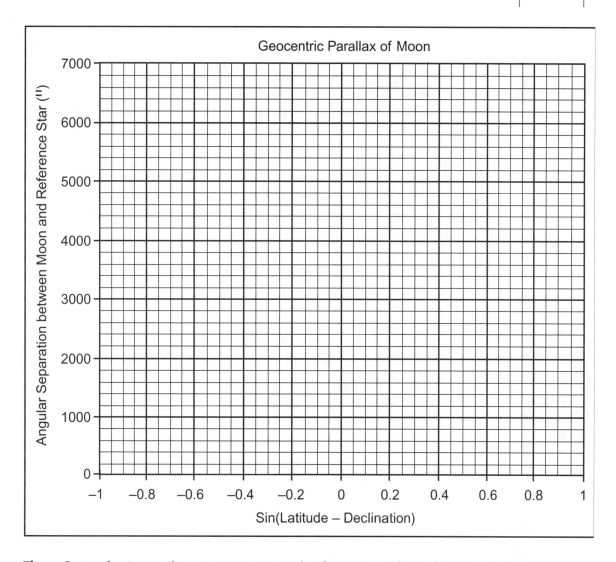

Figure 3. Angular Separation Between Moon and Reference Star from Different Latitudes

The absolute value of the slope of your graph reflects the rate at which the angular separation between the Moon and the reference star (in arcseconds) changes as the latitude of the viewing location and hence the length of the geocentric baseline changes. In mathematical terms, it is the ratio of the change in the angular separation between the Moon and the reference star (in arcseconds) to the change in the quantity sin(latitude – declination). The latter change is equal to the change in the quantity (AB)/R. Thus, because R is constant,

$$slope = \frac{\Delta \ (angular \ separation)}{\Delta \left[\frac{(AB)}{R} \right]} = \frac{\Delta \ (angular \ separation)}{\left[\frac{\Delta (AB)}{R} \right]} \qquad \text{(Equation 4)}$$

where Δ refers to the change in these quantities.

If we choose two points on the Earth such that the change in AB is equal to R, then the change in the angular separation between the star and the Moon equals the parallax of the Moon when measured using a baseline equal to the radius of the Earth; i.e., the angle θ in Equation 2. In the above equation, the numerator in Equation 4 is therefore θ and the denominator unity, so the slope of the graph (except for a minus sign) is equal to θ. We can then use Equation 2 with X equal to the radius of the Earth, 6378 km, and θ = absolute value of the slope of the graph to find the distance to the Moon.

65. Use the absolute value of the slope of your graph as the value for θ and the radius of the Earth, 6378 km, as the value for X in Equation 2 to determine the distance to the Moon in kilometers from your observations.

66. Open the object contextual menu for the **Moon** and select **Show Info.** Under the **Position in Space** layer of the **Info** pane, find the **Distance from observer** for the Moon at this time on this date.

Question 24. What is the distance to the Moon as determined by its geocentric parallax on this date, January 2, 2007?

Question 25. Compare your result to the value you obtained for the distance to the Moon from the **Info** panel. How well do they agree?

Question 26. a) Is the accuracy of your determination of the distance to the Moon better or worse than the accuracy of the result you obtained in section A of this project? b) Why do you think that this is so?

F. Conclusion

In this project, you have measured the parallax of different objects in the solar system to determine their distances from the Earth. You have repeated classical experiments that established the size of the planetary system and thereby laid the foundation for the distance scale of our universe. In doing these simulations, you have used elements of the scientific method such as estimating the accuracy of measurements and graphing of a series of measurements in order to obtain a better estimate of a particular parameter than would be obtained from a single measurement.

Proper Motion of Stars 18

S tars are not fixed in the sky. They move relative to one another, but this motion is very slow compared to that of the Moon and planets. Thus, stars appear to remain fixed in position over the period of a person's lifetime and thousands of years would pass before we would notice significant changes in the shapes of the familiar constellations. It is only by using telescopes and precise measurement techniques in the comparison of measurements or photographs taken many years apart that we can detect the relative movement of stars.

> Proper motion of star is discussed in Box 17–1 of Freedman, Geller, and Kaufmann, *Universe*, 9th Ed.

In general, a star has a velocity in three-dimensional space known as its **space velocity**. The component of this velocity along the line of sight to the observer is known as the star's **radial velocity**. The component perpendicular to the line of sight is known as its **tangential velocity**. Each of these velocities is expressed in units of linear speed such as km/s. The apparent change in position of a star in the sky is known as its **proper motion** and is measured in units of angle as a function of time, such as arcseconds per year. It is this parameter that is measured by comparison of a star's position against a background of more distant objects from a sequence of photographs or images. Tangential velocity of a star can be determined from proper motion, provided that the distance to the star is known. Radial velocity is determined from the Doppler shift of the star's spectrum. A combination of these velocities provides the star's true velocity in space, its space velocity.

Sir Edmund Halley discovered proper motion in 1718. Halley, for whom Comet Halley is named, compared the positions of stars in his time to their positions 1500 to 2000 years earlier as measured by the ancient Greek astronomers Ptolemy and Hipparchus and realized that the stars Sirius, Aldebaran, and Arcturus had moved slightly relative to the other stars.

The discovery and measurement of proper motion of other stars became far easier after the invention of photography and its application to astronomy in the late nineteenth century. With this technique, it became possible to compare simultaneously the positions of many stars on two photographs taken several years apart.

In 1916, the American astronomer E. E. Barnard discovered the star with the highest known proper motion, a faint star in the constellation Ophiuchus. This star is now known as Barnard's star and is situated 5.94 ly from Earth at the present time. It is a small star, having a mass and a diameter of about one-sixth that of the Sun and an apparent magnitude in our sky of +9.5, too faint to see with the unaided eye. It is traveling toward the solar system and will pass within 3.8 ly of the Sun somewhere around 10,000 AD. For comparison, the nearest known stars to the Sun at the present time are those in the Alpha Centauri triple system, at a distance of about 4.3 ly.

In this project, we will investigate the proper motions of Barnard's star and other stars.

A. The Time-Scale of Proper Motion

Proper motion is the angular change in direction of a star in the sky per year, and this change is very slow for most stars. For a star to show proper motion, at least a part of its physical motion through space must be across our line of sight. It is possible to obtain some idea of proper motion by watching the change in the pattern of stars in familiar constellations over a significant time interval in simulation.

1. Select **Favourites > Observing Projects > Proper Motion > Changing Constellations.**
2. Select **File > Revert** to center the view correctly on the constellation Monoceros.

The view is of the winter constellations. Your observing location is at the position of Earth in space, but this location does not move with the Earth and so is unaffected by Earth's rotation or precession.

3. With the **Time Flow Rate** set at **1 year,** observe the constellations as you **Run time forward** for about 1000 years.

Question 1. As time flows at this rate, do you see the sky change?

4. Click the **Now** button in the toolbar.

Question 2. a) Did the constellations appear to change in the last step? b) What does this suggest about proper motion?

5. Observe the constellations and click the AD field in the **Date** display panel in the toolbar. This moves you back in time approximately 4000 years. Click this field again repeatedly to toggle back and forth across 4000 years of time.
6. Use the **Hand Tool** and **Gaze** buttons to observe all of the constellations in the sky, repeating the process of toggling back and forth from AD to BC in time for each region you observe.

Question 3. Are there any of the astronomical constellations that appear to remain completely unchanged over the time span of about 4000 years?

7. **Stop time.**
8. Change the **Date** to **January 1, 4713** BC.
9. Change the **Time Flow Rate** to **200 years.**
10. Use the **Hand Tool** or **Gaze** buttons to choose a region of the sky to observe.
11. **Run time forward.**
12. To repeat this time-lapse view of different constellations, select **Edit > Undo Time Flow,** change the Gaze direction and click **Run time forward.**

Question 4. Is proper motion a common property of stars?

Question 5. a) If a star in a constellation were moving directly toward or away from your line of sight in the simulation, would its motion affect the appearance of the constellation? b) Is this type of radial motion of a star the same as proper motion?

Question 6. A star appears to remain fixed over thousands of years of observation. Can you say for certain that this star is stationary in space?

B. Proper Motion of Barnard's Star

The star with the highest known proper motion is Barnard's star.

13. Select **Favourites > Observing Projects > Proper Motion > Motion of Barnard's Star.**

This view is from the north pole of Earth, in the year 497 AD. The view is centered on reference star TYC421-2053-1 with a field of view of about 6° wide and 4° high.

14. With the **Time Flow Rate at 1 year, Run time forward.**

Because the view is locked on to the reference star TYC421-2053-1, this star and most of its neighbors appear to remain fixed as time advances. However, you will observe three things: First, Barnard's star moves very rapidly across the sky, demonstrating a very high proper motion as it passes almost directly through the position of the reference star sometime in 1107 AD; second, in this small region of the sky spanning a few degrees in width, several other faint stars show noticeable movement with respect to their neighbors over a long time period; and third, the whole sky appears to rotate slowly because of the precession of the Earth's spin axis.

15. Select **Edit > Undo Time Flow** and then click **Run time forward** to see the animation again.

Question 7. Approximately how many years does it take for Barnard's Star to move through the approximately 4° height of the field of view?

16. Select **Favourites > Observing Projects > Proper Motion > Barnard's Star.**

This magnified view shows the position of Barnard's star in the year 1107 AD, the point in time when Barnard's star appears to be very close to the reference star TYC421-2053-1, as seen from Earth.

17. **Step time forward** one step of **10 years.**
18. Use the **Angular Measurement Tool** to measure the angular separation between Barnard's star and the reference star TYC421-2053-1. Record this measurement in Data Table 1 below.
19. Repeat the previous two steps until Barnard's star moves out of the view. Convert your measurements to arcseconds by multiplying the number of arcminutes by 60 and adding the product to the number of arcseconds in the measurement.

Data Table 1. Proper Motion of Barnard's Star

Time Interval (Years)	Angular Motion ' "		Angular Motion (")
10			
20			
30			
40			
50			
60			

20. Plot your observations of the proper motion of Barnard's star on the graph template in Figure 1 below.
21. Attempt to draw a straight line through your data points.

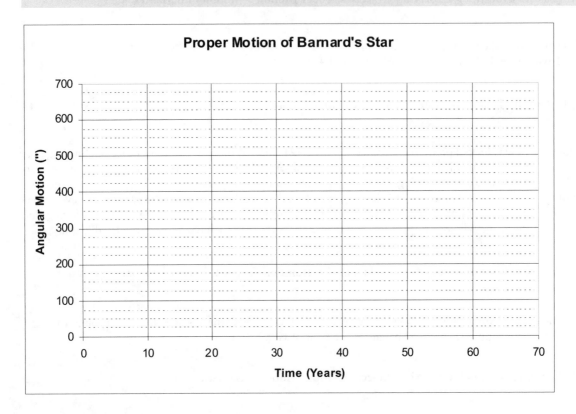

Figure 1. Graph Template for Plotting the Proper Motion of Barnard's star

Question 8. What is the distribution of the data points on your graph?

Notice that the graph in Figure 1 plots angular motion against time. The slope of the line through your data points, the ratio of its rise along the vertical axis over its run along the horizontal axis, is its angular motion in arcseconds divided by the time in years and is therefore the star's proper motion in arcseconds per year.

22. Calculate the slope of the line through the data points on the graph.

Question 9. What is the proper motion of Barnard's star?

Question 10. Does Barnard's star move at a steady rate over time, or does it speed up and slow down occasionally?

Question 11. Toward which direction does Barnard's star move?

23. **Step time backward** until Barnard's star comes back into the view.
24. Position the **Hand Tool** over the image of Barnard's star, open the **Object Contextual Menu** and select **Show Info**.
25. In the **Info** pane, under the **Position in Space** layer, note the **Distance from observer** of Barnard's Star.

Question 12. What is the speed of Barnard's star across our line of sight in this simulation, expressed in AU/yr. [*Hint:* One light year = 63,240 AU. Use this conversion factor with the small angle formula $d = \mu \times R / 206265$, where d = the linear distance that the star moves in a year, μ is the proper motion of the star in arcseconds per year, and R is the distance to the star. The formula then becomes $d = 0.31 \times \mu \times R$ with d expressed in AU/yr.]

As discussed earlier, proper motion of a star is a measure of the component of its motion across our line of sight. Any motion directly toward or away from us will not show up as proper motion.

Question 13. Assuming that Barnard's star is in fact moving directly across our line of sight and has no radial motion toward or away from our observing location, how long would it take Barnard's star to cover the distance between the Sun and Neptune, a total distance of about 30.1 AU?

A complete description of the proper motion of Barnard's star requires that we also specify a **direction** for this motion in the sky. The direction of this proper motion is measured with respect to north and is expressed as position angle (PA), the angle in degrees from north, measured in an eastward direction. For example, proper motion directly toward the east would have a PA = 90°, and a motion toward the southwest would be expressed as proper motion with a PA = 225°. We can use the right ascension-declination coordinate system to provide a north-south reference line against which to measure the direction of the proper motion of Barnard's star.

26. Open the **Object Contextual Menu** over Barnard's star and select **Centre**.
27. Change the **Date** in the toolbar to **April 3, 2004** AD, in order to measure the direction of this motion with respect to the coordinate system for recent times.
28. **Zoom In** to a field of view of about 10' wide.
29. Select **View > Celestial Grid** from the menu.

The view is oriented with north at the top and east to the left. The celestial grid indicates lines of declination horizontally and lines of right ascension vertically. The date has been chosen such that Barnard's star is adjacent to and at the same declination as the star TYC425-262-1. You can use this star as a reference to measure the position angle of the proper motion as Barnard's star moves over a period of 10 years. Figure 2, below, illustrates the geometry of the measurements you will make to determine this angle.

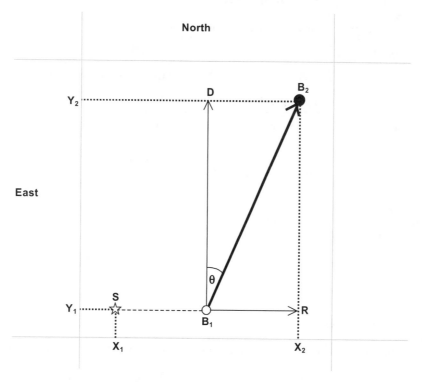

Figure 2. Geometry for Measuring the Direction of the Proper Motion of Barnard's Star

In the current view, Barnard's star is at position B_1, indicated by the open circle, at the same declination as the reference star, S, but separated from this star in right ascension by the angular distance SB_1, indicated by a dashed line. We assume that this reference star has negligible proper motion compared to that of Barnard's star and can therefore be used as a fixed reference for the current position of Barnard's star. (You can verify this using the **Object Contextual Menu** to **Show Info** for this star and find its proper motion in RA and Dec under the **Position in Space** layer of the **Info** pane.)

30. Barnard's star and the reference star are at the same declination in this initial view, so the reference star marks the initial north-south position of Barnard's star. To account for the offset in right ascension, use the **Hand Tool** to measure horizontal angular separation between the reference star and Barnard's star (SB_1 in Figure 2). Record this value, in arcseconds, in the row labeled **Right Ascension Offset, SB_1** in Data Table 2 below.

31. Use the **Object Contextual Menu** to **Centre** the view on to the reference star, TYC425-262-1.

32. **Step time forward** by one interval of **10 years** to **April 3, 2014** AD.

On this date, the proper motion of Barnard's star has carried it to position B_2 in Figure 2. The heavy arrow from B_1 to B_2 in Figure 2 indicates the direction of the star's proper motion. This line makes an angle θ with respect to the north-south direction in the view. This motion can be broken down into components in the vertical (declination) direction and the horizontal (right ascension) direction. Figure 2 shows these components as arrows B_1D and B_1R, respectively. You can use the celestial grid lines in the current view (the light grey lines in Figure 2) and the position of the reference star to measure the angular size of these components of the proper motion of Barnard's star. While these are spherical angles, measured in a spherical coordinate system on the celestial sphere, they are very small. Thus, you can assume that they are linear measurements in the calculation of the angle of the motion of Barnard's star to the coordinate grid.

Data Table 2. Orthogonal Components of the Proper Motion of Barnard's Star over Ten Years

Declination Component of Proper Motion of Barnard's Star	
Angular Distance from Barnard's Star to Declination Line, B_2X_2 (")	
– Angular Distance from Reference Star to Declination Line, SX_1 (")	
= Declination Component of Proper Motion (")	
Right Ascension Component of Proper Motion of Barnard's Star	
Angular Distance from Barnard's Star to RA Line, B_2Y_2 (")	
– Angular Distance from Reference Star to RA Line, SY_1 (")	
– Right Ascension Offset, SB_1 (")	
= Right Ascension Component of Proper Motion (")	

The **vertical** (declination) component of the proper motion of Barnard's star over the 10-year interval is the distance B_1D in Figure 2. You can see from this figure that

$$B_1D = B_2R$$
$$= B_2X_2 - RX_2,$$

and, since $RX_2 = SX_1$,

$$B_1D = B_2X_2 - SX_1$$

You can measure these two distances and calculate this vertical component of Barnard's star's motion.

33. Measure the vertical angular distance from Barnard's star to the first horizontal declination line on the opposite side of the reference star from Barnard's star. Convert this angular distance to arcseconds and enter this value as **Angular Distance from Barnard's Star to Declination Line, B_2X_2** in Data Table 2.

34. Measure the vertical angular distance between the reference star and the same declination line that you used in the previous step. Enter this distance, in arcseconds, as **Angular Distance from Reference Star to Declination Line, SX_1** in Data Table 2.

35. The declination (vertical) component of the proper motion of Barnard's star is $B_1D = B_2X_2 - SX_1$. Calculate this difference and record the result, in arcseconds, in the row labeled **Declination Component of Proper Motion** in Data Table 2.

The **horizontal** (right ascension) component of the proper motion of Barnard's star is B_1R. Again referring to Figure 2,

$$B_1R = RY_1 - SY_1 - SB_1$$

and $RY_1 = B_2Y_2$.

Thus,

$$B_1R = B_2Y_2 - SY_1 - SB_1$$

You have already measured SB_1, from Barnard's star's position 10 years earlier. To determine B_1R, therefore, you need to measure B_2Y_2 and SY_1.

36. Measure the horizontal angular distance from Barnard's star to the first vertical grid line on the opposite side of the reference star (B_2Y_2 in Figure 2). Convert this measurement to arcseconds and record it in the row labeled **Angular Distance from Barnard's Star to RA Line, B_2Y_2** in Data Table 2.

37. Measure the horizontal angular distance from the reference star to the same vertical grid line you used in the previous step. Convert this measurement to arcseconds and record it in the row labeled **Angular Distance from Reference Star to RA Line, SY_1** in Data Table 2.

38. From these measurements, calculate $B_1R = B_2Y_2 - SY_1 - SB1$ and record the result as the **Right Ascension Component of Proper Motion** in Data Table 2.

You can now calculate the position angle of the proper motion of Barnard's star from these measurements. The tangent of the angle between the track of Barnard's star, B_1B_2, and the north direction is the ratio of the right ascension component to the declination component of the proper motion. From Figure 2:

$$\tan(\theta) = B_1R / B_1D$$

39. Calculate the angle θ from the values recorded in Data Table 2 by taking the arc tangent of the ratio of the components of the proper motion of Barnard's star.

40. Since the position angle of proper motion is defined from north, moving eastward, you will need to consider whether this position angle is θ, or $(360 - \theta)$.

Question 14. What was the position angle of the proper motion of Barnard's star in 2004 AD?

You can compare your measured values with those quoted in *Starry Night Enthusiast*™.

41. Open the **Object Contextual Menu** over Barnard's star and select **Show Info** to display information on this star.

42. Open the **Position in Space** layer and note down the values of **Proper Motion RA, P_{RA}** and **Proper Motion Dec, P_d**.

The size and direction of the proper motion of this star can be calculated from these components, since

$$P^2 = (P_{RA})^2 + (P_d)^2,$$

and

$$\tan(\theta) = (P_d/P_{RA}),$$

with θ measured from north through east.

Question 15. a) What is the quoted value for the proper motion of Barnard's star? b) What is the position angle of the direction of Barnard's star's motion, represented as an angle measured from the north direction through east? c) How do your measurements above compare to these listed values?

C. The Proper Motion of Other Stars

The proper motion of other stars is so much slower than that of Barnard's star that we will need a much longer time step in order to see this motion clearly on the screen. We can watch sky motions over the full time span available in *Starry Night Enthusiast™*. If we observe these motions from Earth, the precession of the spin axis causes the whole sky to appear to rotate and this motion of the sky partially masks the effect of proper motion. We can remove this rotation by positioning ourselves at a fixed position in space at a point on the Earth's orbit.

43. Select **Favourites > Observing Projects > Proper Motion > Other Stars.**

The view is of the stars in a region of sky about 23° wide and 16° high near the constellation Hercules in the year 4713 BC.

44. With the **Time Flow Rate** at 100 years, **Run time forward.** When *Starry Night Enthusiast™* reaches its limiting date of 10,000 AD select **Edit/Undo Time Flow** from the menu. Repeat this step several times and watch the proper motion of the stars over nearly 15,000 years of time.

Question 16. Do all of the stars move in the same direction or do different stars move in different directions?

Question 17. Do all of the stars move with the same speed?

Question 18. Which star moves with the highest proper motion in this simulation?

Question 19. Which star moves in quickly from the lower left corner of the view at about 5000 AD?

D. Relationship between Proper Motion and Distance

Suppose you are standing near to a street, looking directly north at the sky. A jetliner that is traveling at about 600 km/hr toward the west comes into your field of view. Without shifting your gaze from the north, you watch as the jet slowly crosses your line of sight. In the meantime, a car moving at about 50 km/hr comes down the street, also traveling west. You will note that the jet is still visible long after the car disappears from your field of view. This is a parallax effect. The car travels much more slowly than the jet, but it is much closer to you and does not need to travel very far in order to travel completely across your field of vision. As you gaze north you might be able to perceive distances about a half block to the east and the west. At the distance of the jet, however, your field of vision spans many kilometers.

In this section, you will explore the statistical relationship between observed proper motion of a star and its distance from the Earth, using a small sample of stars within a particular field of view in one region of the sky.

45. Select **File > Revert.**

46. Select **File > Preferences** to open this dialog and set the **Cursor Tracking/HUD** options to show the **Name, Object type** and **Distance from observer.**

47. **Run time forward** and choose a star that shows high proper motion. When the animation stops at 10,000 AD, use the **Hand Tool** to identify the chosen star and record its **Name** and **Distance from observer** in Data Table 3, below, under the sub-heading **High Proper Motion Stars.** Then select **File > Revert** and repeat this step until you have compiled a list of five stars that show relatively high proper motion.

48. Select **File > Revert.** Then **Run time forward** and choose a star that shows almost no proper motion. Position the **Hand Tool** over the star you chose and record the **Name** and **Distance from observer** for this star from the HUD into Data Table 3 under the subheading **Low Proper Motion Stars.** Repeat this step until you have compiled a list of five stars that show almost no proper motion over the nearly 15,000-year interval.

Data Table 3. Distances to Stars Showing High and Low Proper Motion

Star Name	Distance to Star (ly)
High Proper Motion Stars	
Low Proper Motion Stars	

Question 20. From your survey, what approximate relationship do you see between proper motion of stars in the solar neighborhood and their distances from Earth?

Question 21. What causes this relationship?

49. Exit *Starry Night Enthusiast*™ without saving changes to the files.

E. Conclusions

In this project, you have observed and measured the proper motion of our Sun's fastest-moving neighbor, Barnard's star, as it moves through our sky. You have also been able to catch a glimpse of the complex motions of a small sample of stars near to our Sun as they move in space over many centuries.

The Hertzsprung-Russell Diagram

19

The study of stars occupies a very important place in the overall study of our universe. This is at least partly because the source of most of the energy that we need for our survival on this Earth comes directly or indirectly from one such star, our Sun!

> Chapter 17 of Freedman, Geller, and Kaufmann, *Universe*, 9th Ed, discusses stellar properties and classification and the Hertzsprung-Russell Diagram.

We learn about stars by analyzing the light and other forms of radiation, such as ultraviolet light, X rays, and infrared radiation that come to us across vast distances. This radiation originates at the surface of each star or in its atmosphere. The interior of the star, where most of its mass resides and where all its energy is generated, is hidden from our view.

When we analyze the light from these outer layers, we find that most stars are very similar to one another in terms of their composition. Almost all stars, at least in their outer layers, are composed primarily of the very simple element hydrogen, with a relatively small fraction of helium and a very much smaller fraction of heavier elements. Astronomers refer to all elements heavier than hydrogen and helium as "metals." In terms of their observed composition, it is only this very small metal abundance that varies significantly from one star to another.

Given this great similarity in their observed composition, it is perhaps surprising that stars can take on such a wide variety of forms, from large to small, bright to faint, and hot to cool. It would seem from this that the composition of the surface layers is not the major factor that determines the overall characteristics of a star. The wide array of forms must arise from differences in the interior structure of stars. Also, the simple fact that stars radiate copious amounts of energy means that they must change with time, i.e., stars must "evolve" from one form into another.

To understand these forms and how they might be related to each other, astronomers more than a century ago began to classify stars into categories. The desire to explain why stars fall into these categories then prompted scholars to construct theoretical models of stars using the laws of physics and known properties of matter, such as nuclear and atomic interactions, the emission and absorption of radiation, and the compression of matter by gravity. The most important test of these models is that they must match the observed properties of actual stars, and present-day theoretical models have developed to the point where they can predict the structure and evolution of stars from birth to death with reasonable accuracy.

We will now consider these properties and their classification in a way that is used by professional astronomers and relate them to various physical properties of stars. These classification schemes were devised early in the history of modern astronomy.

Several easily measured stellar parameters are used in the classification of stars. **Stellar brightness** at certain well-defined colors has been measured using the techniques of photometry, initially using visual observations and then photographic imagery, and finally photoelectric measurements. A star's brightness is related to its energy output and is expressed by astronomers in the logarithmic scale of magnitudes. They are referred to as apparent magnitude, m, when they represent measurements made from Earth, with a suffix denoting the color at which the measurement is taken (e.g., m_B is the magnitude measured in the blue region of the spectrum).

Magnitude values increase as brightness decreases in a way that has more to do with history than utility. Stars were assigned a "magnitude" by early Greek astronomers, the brightest being of "first magnitude," the faintest of "sixth magnitude." Astronomers still use a modern version of this inverted logarithmic scale, even though it is inconvenient compared to discussing actual brightness on a linear scale. In this inverted logarithmic scale, a magnitude increase of 1.0—for example, from magnitude 1.0 to magnitude 2.0—represents a brightness decrease of a factor of 2.512. Thus, a change from first magnitude (bright) to sixth magnitude (faint) is five magnitudes difference, which translates to a factor of $2.512 \times 2.512 \times 2.512 \times 2.512 \times 2.512 = 100$ to match approximately the ratio of the brightness of first to sixth magnitude stars as defined by the early Greek observers.

Photometry at different wavelengths can also be used to define a **stellar surface temperature** if the star is assumed to emit approximately as a black body, an assumption that is reasonable for most stars.

Spectroscopy, the detailed examination of the spectrum of stars upon which a complex pattern of dark absorption (and sometimes bright emission) lines appear, also provided a convenient classification method in early studies. Stars with different sets of absorption lines in their spectra were assigned different letters. When it was discovered that the different types of absorption-line spectra were related to stellar surface temperature, these letters were placed in a sequence, O, B, A, F, G, K, and M, relating to decreasing temperature. (A useful key for remembering this important sequence comes from the initials of the words in the sentence "Oh, Be A Fine Girl, Kiss Me!")

Measurement of distances to stars, for a long time a difficult task, began to provide the capability to estimate the absolute brightness of stars by adjusting the apparent brightness for distance via the inverse square law. This led to absolute magnitudes, designated by the capital letter "M," again with a suffix appropriate to the color at which this brightness measure is defined. Absolute magnitude is defined as the magnitude that a star would have if it were at a standard distance of 10 parsecs, or 32.6 light-years, from Earth and therefore represents a measure of star output that is independent of distance from the observer.

These brightness values at certain colors could then be combined by using the overall shape of the black body spectrum of the star derived from the temperature to provide an estimate of the total energy output of the star, its **luminosity**, L. This measure of total energy output of the star, usually represented in terms of the luminosity of the Sun, is an important parameter for comparison with predictions of theoretical stellar models.

Finally, there is a relationship between a star's luminosity, L, temperature, T, and size represented by its radius, R. A black body at a temperature T is known to emit a certain amount of energy per unit area per second, E, that is related to its temperature by $E = \sigma T^4$, where σ is a constant. If we assume that a star behaves like a black body in emitting light, a star of radius R with a surface area of $A = 4\pi R^2$ will emit a total amount of energy per second represented by its luminosity $L = E \times A$, or $L = 4\pi R^2 \sigma T^4$.

Early in the last century, two astronomers, Ejnar Hertzsprung in Denmark and Henry Norris Russell in the United States, began to experiment with ways to represent stars collectively in terms of these observed and derived parameters. They independently hit upon a graphical method that has become the keystone, not only of classification of stars, but also of the study of the evolution of stars with time.

The **Hertzsprung-Russell diagram**, or H-R diagram as it is often called, is a scatter diagram of the luminosity of stars plotted as a function of their surface temperatures. In view of the wide range of these parameters for stars, the luminosity and temperature are usually plotted with non-linear scales. Luminosity is plotted logarithmically along the vertical axis, while the horizontal axis shows either temperature plotted logarithmically, or the spectral letter classifications O, B, A, F, G, K, and M plotted equally spaced. (The latter sequence actually corresponds quite closely to the logarithmic temperature scale.) Other versions of the H-R diagram may denote absolute magnitude of the stars against their spectral classification. All H-R diagrams plot temperature in an inverse direction, with high temperatures on the left and low temperatures on the right.

There are several characteristics of note in this scatter plot of stellar parameters. Stars tend to bunch together in certain areas of the plot whereas other areas are almost devoid of stars. The majority of stars are concentrated in a diagonal sweep across the diagram from high temperature-high luminosity to low temperature-low luminosity (i.e., from upper left to lower right). This region, known as the **main sequence**, is where theoretical models predict that stars will congregate when they are "burning" hydrogen in their interiors in nuclear reactions. The lower edge of the main sequence, known as the Zero Age Main Sequence (ZAMS), designates the line where stars of different mass first begin to burn hydrogen in their cores. This line is thus the locus of points for stars that have reached the time in their lifetimes when pressure generated by the heat of nuclear burning in their cores is sufficient to balance the gravitational force that has been condensing the pre–main-sequence stars.

Few stars are found below the ZAMS. However, those that are there form a class of very hot stars that, despite being very hot, are so small that their luminosity is very small as a consequence. These white dwarf stars represent old and very evolved stars that have shed their outer layers to reveal very small but extremely hot and dense inner cores. These stars are no longer generating energy but are merely emitting light as they cool.

Stars with high luminosities but relatively low temperatures occupy a wide region above the main sequence. Theoretical prediction indicates that the majority of these are stars have consumed all the hydrogen in their cores and have expanded and cooled as a result of internal readjustment but are still burning helium and other elements. These are the red giants.

There are stars with enormous outputs of energy represented by very high luminosities that cannot be very old because they are burning their fuel at a prodigious rate. These are the supergiants, and they can be hot or cool, hence blue or red in color.

The final stellar parameter that is important in this work is the **mass** of the star. This is a parameter that is difficult to measure but vital to this modeling. Stellar mass represents the amount of fuel that is available for nuclear burning. However, it is also found that the larger the initial mass of the star, the hotter the star's core. This leads to more vigorous nuclear fusion. Thus, the larger the mass, the higher the energy output or luminosity. The energy output of a higher-mass star is so great that, despite the larger amount of fuel, it uses up its fuel in a shorter time than does a lower-mass star. Because of this effect, we find a relationship between the mass of a main-sequence star and both its luminosity and its lifetime: The larger the mass of a star, the greater is its energy output and the shorter is its lifetime. This in turn leads to the conclusion that the more massive stars at the high-temperature, high-luminosity end of the main sequence are young and will evolve rapidly away from this position. The mass-luminosity relation, shown as a graph of luminosity versus mass later in this chapter, will be used for determining masses of selected stars from an observed list.

Two further concepts concerning the H-R diagram are interesting. The first concerns regions on this diagram where no stars appear. There are two reasons why this can be so. First, conditions may never be suitable to produce a star with these luminosities and temperatures. Second, a star can possess these values of luminosity and temperature, but its transition through this region in its evolution is so rapid that the probability of finding a star in this position is small. Theoretical modeling of stellar evolution can predict a star's path across the H-R diagram (and thus delineate the regions of the H-R diagram that cannot be occupied by stars) and the speed at which it follows this path. The observed number of stars in the different regions of the H-R diagram therefore provides an important test of the theory.

The second concept concerns the classification of stars in a cluster of stars. A cluster forms from a single interstellar cloud, or even from just part of a single interstellar cloud if it is a large cloud, so we can be reasonably sure of three conditions. First, an interstellar cloud has a relatively uniform composition, so the stars in the cluster form with the same chemical make-up. Second, we can assume that all of these stars formed at about the same time and therefore have the same age. Third, the differences in observed star properties are then caused by differences in the initial mass that came together to form each star. This makes the study of clusters important for the verification of evolutionary models. The massive and highly luminous stars at the top end of the main sequence evolve more rapidly and so leave the main sequence sooner. This means that such stars will be absent in an old cluster. Indeed, the position of the upper end of the remaining main sequence on the H-R diagram of a cluster, the so-called turn-off point, is a good indication of the age of the cluster.

In this project, you will examine and classify a series of chosen stars to build up your own composite H-R diagram. Of course, you cannot measure all the various necessary parameters of these stars. Astronomers have accomplished this over many years and have provided the relevant information that is in the *Starry Night Enthusiast*™ data bank. You can find each of the listed stars with this program and use the **Info** pane to obtain the data for each star in turn.

Thus, you will be able to classify these stars and reach some conclusions about their past and future by comparing their positions on your H-R diagram with the results of stellar evolution modeling.

A. Classification of Stars on an H-R Diagram

The list in Data Table 1 contains many bright stars in our sky, with a wide range of properties. The procedure will be to find each star in turn using *Starry Night Enthusiast*™, retrieve the necessary information from the **Info** pane for each star and tabulate the information in Data Table 1. You can then plot these parameters on the H-R diagram graphical template and classify the stars into the different groups discussed earlier. The parameters that are most easily plotted on this graph are absolute visual magnitude on the vertical axis against temperature on the horizontal axis. The brightness representation is linear but on an inverted scale; the temperature scale is reversed and non-linear, but the intervals of temperature are marked such that points can be plotted easily.

1. Launch *Starry Night Enthusiast*™.

2. Select **Favourites > Observing Projects > HR Diagram > Local Neighborhood Stars.**

3. For each star listed in Data Table 1, open the **Find** pane, type the name of the star exactly as it is written in the data table, and press **Enter.** The view changes to center the chosen star in the main view window. You can speed up this movement by pressing the **Spacebar.**

4. In the **Find** pane, click the menu button to the left of the name of the star in the list and select **Show Info** from the drop-down menu.

5. Expand the **Other Data** layer in the **Info** pane, and copy the values of the star's **Absolute Magnitude (Visual)** and **Temperature** into Data Table 1.

6. If you wish, you can choose other stars at random to fill out your H-R diagram, or add your favourite star to this graph.

White dwarf stars are either too faint or too close to bright companion stars for *Starry Night Enthusiast*™ to display them. To fill in this gap, Data Table 1 includes the parameters of a few representative white dwarf stars for inclusion in your H-R diagram.

Figure 1 is an H-R diagram template that you can use to plot the absolute magnitude of each star against its temperature. (You can print out extra copies of this template. Navigate to the folder **Program Files/Starry Night Enthusiast 5/Sky Data/Extras.** In this folder, you will find this figure in several formats: a Word document, a bitmap, and a pdf file.)

The template includes indications of the regions where different types of stars are found, including supergiants, giants, main-sequence stars, and white dwarfs. Within the main-sequence region, a solid line shows the approximate center of the main-sequence band. The temperature scale along the bottom increases toward the left and is logarithmic. Across the top of the diagram, the spectral classifications are shown. The absolute visual magnitude, a linear scale, is on the left vertical axis. The luminosity scale on the right is logarithmic and indicates a star's luminosity compared to that of the Sun, whose luminosity is one on this scale. Diagonal lines in the background of the graph show a logarithmic grid of stellar radii, normalized to the Sun's radius. These lines represent the relationship between the radius of the star, its luminosity and its temperature described above.

7. Note that each of the stars in Data Table 1 has been assigned an **ID** number. To plot a star on Figure 1, or on a printout of the template, write its ID number inside a small circle indicating its position on the graph. Plot the absolute magnitude of each star in Data Table 1 vertically against the temperature horizontally. Note that the absolute magnitude scale is inverted, increasing downward, and the temperature scale is reversed and non-linear.

8. From the star's position on the graph, determine whether it is a white dwarf, main-sequence, giant, or supergiant star and record this in the column labeled "Type of Star" in Data Table 1.

Hertzsprung–Russell Diagram Template

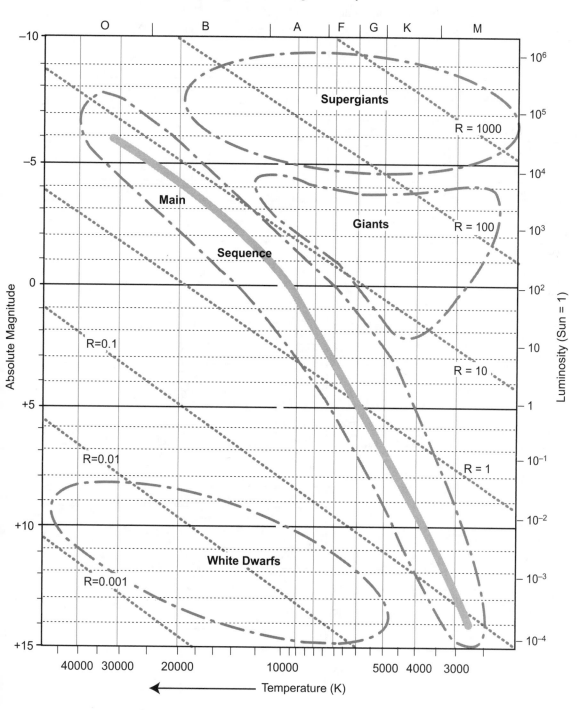

Data Table 1. Absolute Magnitude and Temperature of Selected Stars

ID #	Star	Absolute Magnitude (visual)	Temperature (K)	Type of Star
1	Vega			
2	Aldebaran			
3	Sirius			
4	Capella			
5	Procyon			
6	Rigel			
7	Betelgeuse			
8	Bellatrix			
9	Alhena			
10	Merope			
11	Miaplacidus			
12	Adhara			
13	Fomalhaut			
14	Altair			
15	Beta Pictoris			
16	Alpha2 Centauri			
17	Mu Cassiopeiae			
18	Alphard			
19	Sadr			
20	Altais			
21	Sarin			
22	Theta Lyrae			
23	Spica			
24	Deneb			
25	Canopus			
26	Polaris			
27	Epsilon Eridani			
28	Phecda			
29	Pollux			
30	Antares			
31	Alioth			
32	Alpha Lupi			
33	Nunki			
34	Acrux			
35	Shaula			
36	Almaaz			
37	Sirius B	+11.5	26,000	White Dwarf
38	40 Eridani B	+11.0	15,000	White Dwarf
39	Procyon B	+ 13.0	8,200	White Dwarf

Question 1. Are there regions on your H-R diagram that appear to be devoid of stars? If so, where are they (e.g., high-temperature/high-luminosity, low-temperature/high-luminosity, etc.)?

Question 2. Are these regions also devoid of stars in more extensive H-R diagrams displayed in your textbook or other astronomy texts? (See, for example, Figure 17–15(a) in Freedman and Kaufmann, *Universe*, 8th ed.)

Question 3. Based on the three white dwarf stars, what is the luminosity of a typical white dwarf star in terms of the Sun's luminosity?

Question 4. What are the radii of the three white dwarf stars?

Question 5. To which spectral classifications might a white dwarf star belong?

Question 6. In which classification group—white dwarf, main-sequence, giant, or supergiant—is the star Deneb (ID# 24)?

Question 7. To which spectral class does the star Deneb belong?

Question 8. a) Which is the most luminous star in the list? b) To which spectral class does it belong? c) To which group (i.e., white dwarf, main-sequence, giant, supergiant) does it belong? d) Approximately how much more luminous is this star than the Sun? e) How does this star's radius compare to that of the Sun? f) What is the ratio of the surface temperature of this star compared to that of the Sun, 5800 K?

Question 9. a) Which star in the list has the coolest surface temperature? b) To which spectral class does it belong? c) To which group (white dwarf, main-sequence, giant, or supergiant) does it belong?

Question 10. a) Which star in the list has the hottest surface temperature? b) To which spectral class does it belong? c) To which group does this star belong?

Question 11. What is the approximate range of radii of stars in the main sequence compared to the size of the Sun?

Question 12. a) Could a main-sequence star with the radius of the Sun achieve the luminosity of Deneb? b) For a star to have the same luminosity as Deneb and still be on the main sequence, what would its radius and temperature have to be?

Question 13. a) Of the stars in the list, which is the closest in size and luminosity to the Sun? b) Do you think that the Sun is much hotter or cooler than this star? Why?

Question 14. What is the approximate maximum surface temperature that a star the size of the Sun could have and still be on the main sequence? What is the approximate minimum temperature that a star the size of the Sun could have and still be on the main sequence?

Question 15. Approximately how much larger would the radius of a star with the same surface temperature as the Sun (5800 K) have to be before it could be classed as a) a giant star, and b) a supergiant star?

Question 16. Approximately how much hotter would a star the size of the Sun need to be in order to reach the same luminosity as a) a giant star, and b) a supergiant star?

B. Masses of Main-Sequence Stars

In the previous section, you were able to classify a list of bright stars on the H-R diagram and identify several main sequence stars among them. In this section, you will use the relationship between the luminosity and mass of main-sequence stars to determine the masses of several stars.

The mass is an important physical parameter of a star but is difficult to measure directly. In fact, the mass of a star can be measured directly only if it is a member of a binary system in which the star's gravitational influence on its companion star can be observed. This can be done only for a select few nearby stars where the properties of the binary can be measured. Thus, the mass-luminosity relation for main-sequence stars, verified for a few stars by direct measurement, is very important in providing estimates of mass for a much wider range of stars.

The theoretical basis for this relationship comes from the idea that the more massive the star, the more intensely the nuclear furnace will burn the hydrogen in its interior, producing a higher energy output from the star, i.e., a higher luminosity. The graph shown in Figure 2 below is a summary of main-sequence stars with a range of masses between 0.1 and 20 solar masses. Densities and temperatures of hydrogen in the interiors of stars with masses lower than about 0.1 solar masses will never be sufficient to trigger nuclear reactions, and such stars will merely emit energy that has been generated by the contraction of the gas by gravity. Because of the absence of nuclear reactions, these objects are not true stars, and are usually called brown dwarfs. Above a higher mass limit of about 80 to 100 solar masses, stars become extremely unstable, the nuclear furnace burning so vigorously that they rapidly reach an explosive and destructive stage.

> The Mass-Luminosity relation is discussed in Section 17–9 and Figure 17–21 in Freedman, Geller, and Kaufmann, *Universe*, 9th Ed.

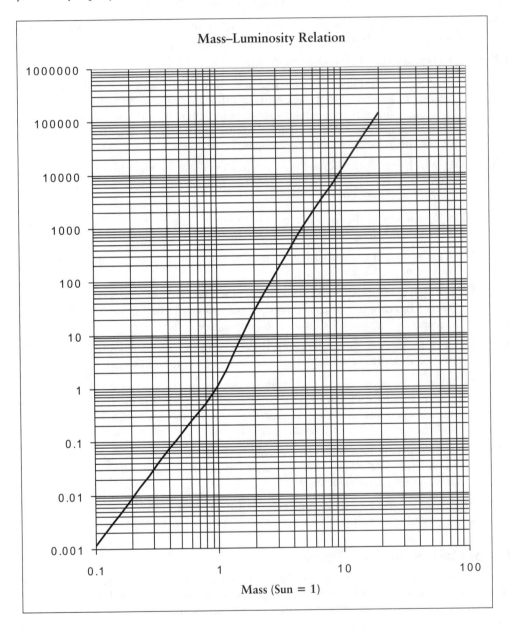

Figure 2. Mass-Luminosity Relation

Data Table 2 below contains a sample of some of the main-sequence stars that you plotted on the H-R diagram in the previous section, subdivided into high-, medium-, and low-temperature stars. You can use Figure 2 to determine their masses. Notwithstanding the fact that both axes in this graph are logarithmic in scale, you should be able to obtain a reasonable estimate of the mass of each of these stars.

9. For each of the stars listed in Data Table 2, use the **Find** pane to locate them in *Starry Night Enthusiast*™ and then click the menu button to the left of the star's name in the **Find** pane list and select the **Show Info** option from the drop-down menu. Under the **Other Data** layer of the Info pane, find the value for the star's **Luminosity** in terms of the luminosity of the Sun, and copy this value into Data Table 2.

10. To find the mass of each of these stars, use the value of its **Luminosity** on the vertical axis of Figure 2 and read off the associated value of **stellar mass,** in units of the solar mass, along the horizontal axis of Figure 2, taking careful note of the fact that both axes of this graph have logarithmic scales. Enter this mass in the appropriate box of Data Table 2.

Data Table 2. Masses of Some Main-sequence Stars

Star	Luminosity (Sun = 1)	Mass (Sun = 1)
Low Temperature		
Alpha2 Centauri		
Mu Cassiopeiae		
Medium Temperature		
Procyon		
Fomalhaut		
Altair		
Beta Pictoris		
High Temperature		
Bellatrix		
Spica		

Question 17. How does the range of masses compare to the range of luminosities in Data Table 2?

The mass-luminosity relationship for main-sequence stars can be represented approximately by a power law, in which luminosity is proportional to mass to a high power, $L = k\,M^{3.6}$, where k is a constant. This equation shows that stars with large masses will have very much larger luminosities than their lower-mass counterpart.

C. Average Density of Stars

Now that you have determined the masses of several main-sequence stars you can estimate their radii from their positions on the H-R diagram and use these values to determine the average density of the gases in these stars and compare it to a typical density on Earth, such as that of water, whose density is 1000 kg/m³.

The average density, ρ_S, of a star is the ratio of its mass, M_S, to its volume, V_S, which is given by $4\pi R_S^3/3$, where R_S is the radius of the star. Therefore,

$$\rho_S = \frac{M_S}{4\pi R_S^3/3} = \frac{3M_S}{4\pi R_S^3} \qquad \text{(Equation 1)}$$

Since the values you have for the mass, M_S, and the radius, R_S, of these stars are given in terms of solar mass, M_{Sun}, and solar radius, R_{Sun}, respectively, you will need to do some conversions to obtain a value for the average density of these stars in terms of kg/m³ to compare your calculated values to the density of water. To make these conversions, you need to multiply the value of the mass of the star in terms of solar masses by the mass of the Sun and, similarly, you need to multiply the radius of the star in terms of solar radii, by the radius of the Sun. Therefore,

$$\rho = \frac{3M_S M_{Sun}}{4\pi(R_S R_{Sun})^3} = \left[\frac{M_S}{R_S^3}\right]\left[\frac{3M_{Sun}}{4\pi R_{Sun}^3}\right] \qquad \text{(Equation 2)}$$

The term on the right in the above equation is simply the average density of the Sun, ρ_{Sun}. Therefore,

$$\rho_S = \left[\frac{M_S}{R_S^3}\right]\rho_{Sun} \qquad \text{(Equation 3)}$$

Using values of 1.99×10^{30} kg for its mass, and 6.96×10^8 meters for its radius, the average density of the Sun is $\rho_{Sun} = 1410$ kg/m³. Therefore, the average density in kg/m³ of a star whose mass, M_S, and radius, R_S, are given in terms of the solar mass and radius is:

$$\rho_S = 1410\, \frac{M_S}{R_S^3}\ \text{kg/m}^3 \qquad \text{(Equation 4)}$$

To compare the average density of the star to the density of water, divide the result of Equation 4 by 1000 kg/m³.

11. For each of the main-sequence stars listed in Data Table 3 below, copy the values of mass of the star, M_S, in terms of the solar mass, for each of these stars from Data Table 2 into Data Table 3.

12. Use *Starry Night Enthusiast*™ to find the radius of each star in Data Table 3 in terms of the solar radius. Open the **Find** pane, type the name of the star into the search box and press **Enter**. Then click the menu button to the left of the star's name in the **Find** list and select **Show Info** from the drop-down menu. Under the **Other Data** layer, you will find the value for the **Radius** of the star in terms of the radius of the Sun. Copy this value into Data Table 3 below.

13. Finally, use Equation 4 to calculate the average densities of these stars by multiplying the ratio M_S/R_S^3 by 1410. Divide this result by 1000 to determine the stars' average densities in terms of the density of water.

Data Table 3. Approximate Densities of Some Main-sequence Stars

Star	Mass, M_S (Sun = 1)	Radius, R_S (Sun = 1)	Average Density, ρ_S (kg/m³)	Average Density (Water = 1)
Low Temperature				
Alpha2 Centauri				
Mu Cassiopeiae				
Medium Temperature				
Procyon				
Fomalhaut				
Altair				
Beta Pictoris				
High Temperature				
Bellatrix				
Spica				

Question 18. Which of the stars in Data Table 3 have average densities greater than water?

Question 19. a) Which star from the list in Data Table 3 has the lowest average density? b) Which star from the list in Data Table 3 has the highest average density? c) What is the ratio of highest to lowest average density in this selection of stars?

As you can see from Data Table 3, there is a significant range of average density among main-sequence stars. Of course, the density of a star varies enormously over its volume, from extremely low density at the outer reaches of the star's atmosphere to extremely high density in its central core.

It is interesting to calculate the density of a white dwarf star. White dwarfs are the burned out cores of low-mass stars and consist of degenerate matter condensed to very high densities. While main-sequence stars show a relationship between mass and luminosity, white dwarf stars show a relationship between mass and radius, with mass increasing as the radius of the star decreases.

Question 20. The white dwarf 40 Eridani B has a radius of 0.014 solar radii and a mass of 0.43 times that of the Sun. What is the average density of this star, in kg/m³? b) How much greater is this than the density of water? c) The volume of a teaspoon is about 5 cubic centimeters. What is the mass of a teaspoon of matter from 40 Eridani B?

D. Variable Stars

Many stars are found to vary in brightness periodically with a wide range of oscillation periods. Some of these stars vary in brightness with precise regularity, while others are much less regular. This variation in brightness is accompanied by coincident changes in temperature and size. When placed upon the H-R diagram, these stars are found to occupy positions that are related to their period. For example, a range of stars with regular periods of between 1 and 100 days are known as Cepheid variables after the first star of this type discovered, δ Cephei. They occupy an almost vertical band above the main sequence known as the instability strip, extending across a range of luminosities through the yellow giant and supergiant regions. A similar class, the RR Lyrae stars, with more restricted periods of less than one day, occupies a small region with a narrow range of luminosity at the bottom of the instability strip.

An important relationship has been established between the period of Cepheid variables and their luminosity. Thus, merely recognizing that a star is a Cepheid variable and measuring its period establishes its intrinsic brightness. When combined with a measurement of its apparent brightness, this information can be used to provide a measure of its distance from the Sun. Because Cepheid variables are sufficiently bright that they can be identified in nearby galaxies, this method of distance measurement is one crucial rung in the so-called distance ladder, allowing astronomers to determine distances over vast ranges and estimate the overall scale of the universe.

Other stars with much longer and somewhat variable periods, the long-period variables, LPV, occupy a region on the cooler side of the H-R diagram, again extending over a wide range of luminosities.

Data Table 4 contains a list of several different variable stars that can be found using *Starry Night Enthusiast*™.

14. For each star in Data Table 4, type its name in the search box at the top of the **Find** pane and press **Enter.** Then click the menu button for the star in the **Find** list and select **Show Info.** Under the **Other Data** layer, find the **Absolute Magnitude (Visual)** and the **Temperature** and copy these values into Data Table 4.

15. Using the ID letters provided in Data Table 4, plot these stars on the H-R diagram that you constructed in section A of this project. You might want to use a different color to plot these variable stars or use a fresh template and compare your two H-R diagrams.

Data Table 4. Absolute Magnitude and Temperature of Some Variable Stars

ID Letter	Star	Period (days)	Absolute Magnitude	Temperature (K)
Cepheid Variable Stars				
A	T Monocerotis	27.02		
B	Zeta Geminorum	10.15		
C	U Vulpeculae	7.99		
D	Eta Aquilae	7.18		
E	T Vulpeculae	4.44		
RR Lyrae Variable Stars				
F	RR Lyrae	0.567		
Long PeriodVariable Stars				
G	Omicron Ceti	358		
H	R Aurigae	245		
I	U Orionis	307		
J	V Monocerotis	302		
K	T Hydrae	426		
L	R Leonis	407		
M	R Virginis	378		

Question 21. Which of the stars in Data Table 4 has a) the largest radius? b) the highest luminosity? c) the smallest radius? d) the least luminosity?

Question 22. Through which regions (e.g., white dwarf, main-sequence, giants, supergiants) of the H-R diagram does the instability strip represented by the variable stars in Data Table 4 extend?

Question 23. To which of the spectral classes (e.g., O, B, A, F, G, K, M) do the variable stars in Data Table 4 mostly belong?

Question 24. Which of the stars in Data Table 4 is closest to the centerline of the main sequence?

E. Conclusion

In this project, you have emulated professional astronomers in plotting an H-R diagram for a selected set of typical stars, but without having to measure spectra, star brightness, or star distances, as an astronomer would. However, you have the satisfaction that you have been able to use the best available data set in the world for this type of investigation, the Hipparchos database incorporated into *Starry Night Enthusiast™*. You have located a few representative white dwarf stars and variable stars for comparison with other classes of stars on this important classification tool, the Hertzsprung-Russell diagram.

The Distance Ladder   20

I t has been less than 100 years since astronomers came to appreciate the immense size of our universe. As recently as 1920, the astronomer Harlow Shapley argued in a celebrated debate with Heber Curtis of Lick Observatory that the Milky Way encompassed the entire universe. Prior to this debate, Shapley had used the distribution of globular clusters in the sky to deduce the position of the Sun within the Milky Way and to provide a better insight into the size of this galaxy. However, he also concluded erroneously that "spiral nebulae" were within the Milky Way and argued this point forcefully in the debate. It was only 3 years later, in 1923, that Edwin Hubble, working at Mount Wilson Observatory, found very strong evidence to prove Shapley wrong. Hubble discovered that one of these "spiral nebulae," the Andromeda Nebula, was in fact a completely separate galaxy far removed from our own Milky Way. This initial discovery was soon followed by the identification of many other galaxies, or "separate worlds," to use Curtis's phrase. In the 80 or so years since Hubble's discovery, astronomers have found evidence that the universe contains billions of other galaxies, and have developed techniques for estimating their distances. In this way, astronomers have systematically probed the limits of the observable universe, discovering objects at distances so vast that it has taken billions of years for their light to reach us. Our understanding of the size and evolution of the universe has been profoundly altered by these discoveries.

Astronomers use various methods to measure distances in the cosmos. Each of these methods is useful over a particular range. Fortunately, these ranges overlap to some extent, allowing astronomers to calibrate farther-reaching techniques against more reliable closer-range methods. In this way, a so-called distance ladder has been assembled, each rung of the ladder representing a technique that extends the range of measurable distances in the universe.

In this project, you will "climb" the distance ladder. You will ascend each "rung" by making observations that use or simulate the distance measuring technique appropriate to that rung.

A. Parsecs and Light-years

The parsec is a unit of distance based on the first rung of the distance ladder, the parallax effect. As discussed in a previous project, parallax is the apparent shift in position of a relatively nearby object against a more distant background, caused by a change in an observer's position. Closer objects have a greater parallax than distant objects. The parallax effect also depends on the distance between the two viewing points, the observational baseline. In the previous project devoted to parallax, you learned about geocentric and diurnal parallax, based on an observational baseline limited to the diameter of the Earth. With this baseline you could measure the distances to objects in the solar system, a range of several tens of AU. Measuring the parallax of more distant objects such as stars requires a longer baseline than the Earth's diameter. The baseline used to determine stellar distances is the diameter of the Earth's orbit. The angular shift of a nearby star is measured against the background of more distant stars when observed from the Earth at two times that are separated by 6 months, during which time the Earth has moved to the opposite position in its orbit about the Sun, a distance of 2 AU. This method is called **stellar** or **heliocentric parallax**.

A parsec is defined as the distance at which a star as seen from the Earth shows an annual *par*allax of 1 arc*second*. The annual parallax of a star is equal to one-half of the angular shift of its position due to parallax when measured from Earth at 6-month intervals. With the parsec defined this way, the formula for calculating the distance, *d*, of an object that shows an annual parallax of *p* arcseconds is

$$d = \frac{1}{p}$$

Seen from another perspective, a parsec is equivalent to the distance from which the average radius of the Earth's orbit, 1 AU, would span an angular size of 1 arcsecond.

1. Launch *Starry Night Enthusiast*™.

2. Open the **Preferences** dialog and change the **Cursor Tracking (HUD)** options to display the **Name** and **Distance from observer.**

3. Select **Favourites > Observing Projects > Distance Ladder > Parsec.**

The view is from a location in space toward the north ecliptic pole so that the Sun is centered and the plane of the Earth's orbit is face-on. You should note that the field of view has been zoomed in to only 4″ wide and 3″ high.

4. Measure the angular separation between the Sun and the Earth.

Question 1. What is the physical distance in AU between the Sun and the Earth in the view? [*Hint:* The **Angular Measurement Tool** tells you the distance between the two objects you are measuring.]

Question 2. What is the angular separation between the Sun and the Earth in the view?

Question 3. What is the distance of the viewing location from the Sun in parsecs?

Another unit used for measuring vast astronomical distances is the light-year, the distance that light, moving with a speed of 3×10^8 m/s, travels in 1 year. One parsec is equal to 3.26 light-years, so the parsec is in fact a larger unit of distance than the light-year.

Question 4. What would be the observed parallax angle of a star 1 light-year away from Earth?

With this introduction to the units appropriate to the vast distances you will measure in the following sections of this project, you are now ready to climb the first rung of the distance ladder.

B. Stellar Parallax

Despite the larger baseline used in stellar parallax observations, even the nearest stars have parallax angles of less than 1 arcsecond. Stellar parallax is thus a very subtle and difficult measurement to make. It was not until 1838 that Wilhelm Friederich Bessel made the first successful measurement of stellar parallax. In recent years, astronomers have been able to use modern ground-based telescopes to measure parallax angles of 0.01 arcseconds. The most recent and major advance in this important field of astrometry has been the satellite *Hipparcos*, whose name is an acronym for High Precision Parallax Collecting Satellite. Despite being placed in a wrong orbit, this satellite revolutionized our knowledge of star distances and positions by measuring the parallaxes of over 100,000 stars to an accuracy of 0.001 arcseconds in the early 1990s. Star positions and properties derived from this space mission are included in the *Hipparcos* database and its companion, the *Tycho* database, and these databases have been incorporated into *Starry Night Enthusiast*™, hence the HIP and TYC prefixes associated with stars in this program.

Thus, parallax measurements using these techniques have been able to determine distances of objects out to about 1000 parsecs or 3300 light-years from the Sun.

Question 5. Assuming that a parallax angle (i.e., one-half of the displacement angle) of 0.01 arc-seconds is the minimum that can be reliably measured using telescopes on the Earth, what is the maximum distance (in parsecs) that can be determined from parallax observations?

Question 6. If *Hipparcos* measured a parallax angle of a particular star to be 0.004 arcseconds, what is the distance of this star from the Earth?

In the next sequences you will use *Starry Night Enthusiast*™ to measure the angular shift of some of the stars in the local neighborhood of the Sun. To make these subtle measurements, you will view the change in the position of a star as time progresses through a year from a location attached to the Earth as it orbits the Sun. The chosen viewing location is about 2200 AU above the north pole of the Earth. Above this elevation, *Starry Night Enthusiast*™ disables the display of the proper motions of stars that otherwise would mask stellar parallax. By placing the viewing location at an elevation above the north pole, the observing location moves through a duplicate orbit to that of the Earth with a radius of 1 AU but displaced 2200 AU toward the north celestial pole in space. The observed effect is just as if you were viewing the parallax effect from the Earth.

As the Earth carries the observing location around an orbit of radius 1 AU, a star will appear to describe an ellipse in the sky. If the star is close to the ecliptic pole in the sky, then you are looking up at it and its motion will mimic the (almost) circular motion of the orbit of the Earth. However, if the star is on or near to the ecliptic plane, your motion will be back and forth across this plane and the star will appear to move back and forward in a long ellipse or even, for stars right on the ecliptic plane, a straight line. For each of the stars listed in Data Table 1 below, follow the steps in the remainder of this section to measure its parallax.

Sequence 1

5. Open the **Favourites** pane and select the view that matches the name of the star in Data Table 1 from the **Observing Projects > Distance Ladder** folder beginning with **Alpha2 Centauri**.

The view is from a location elevated about 2200 AU above the north pole of the Earth. The star is labeled near the center of the view.

6. **Zoom In** to a field of view close to or at the limit of **<1″ × <1″** (In several cases, the star may move out of the view at this limit and you will have to expand the field of view to a few arcseconds).

As you zoom in, watch the field of view indicator change in the compass display in the right upper corner of the view window, as it indicates the direction of the star in the sky. Also, as you decrease the field of view and get close to maximum zoom, you will see the star drift away from the center of the view. This is because the gaze direction is centered and locked onto a point in a blank patch of the celestial sphere near to but not at the star.

7. With the **Time Flow Rate** set at **3 days, Run time forward** and watch the parallactic motion of the star over a year or so of time and note the direction of the star's motion (clockwise, CW, or counterclockwise, CCW) and the approximate shape of the motion (circle, ellipse, long ellipse, straight line) in the appropriate boxes of Data Table 1.

As time runs forward, your observing location is being carried along with the Earth in its orbit around the Sun and shows the position of the star changing relative to the fixed point in the infinite distance of space to which the view is locked. This motion is the result of the parallax effect as seen from an orbiting Earth. As explained above, its shape is generally elliptical as a result of the projection of this orbital motion at the position of the star above or below the ecliptic plane.

You can now go ahead and measure the length of the long axis of the ellipse, which is twice the parallax angle for the star, by tracing the position of the star every 10 days over the course of slightly more than a year. Figure 1, below, is a hypothetical example of what you will see as you follow the steps for measuring the parallax of the star. Note that the star's path does not join up exactly on itself because of an additional drift across the sky that is caused by the slow precession of the Earth's spin axis. The diagram illustrates how to remove the effect of the extra drift and make an accurate measurement of the parallax shift from your tracing.

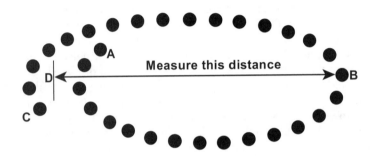

Figure 1. Parallax Motion of an Hypothetical Star

8. Change the **Time Flow Rate** to **10 days.**

9. **Step time backward** or **forward** so that the star is approaching but not yet at one of the ends of the long axis of its motion as in the point marked A in Figure 1.

10. Tape a sheet of transparent plastic or clear tracing paper to the screen and mark the star's position with a soft felt pen or marker.

11. **Step time forward** in 10-day steps and plot the parallax motion of the star on the paper. Plot the star's path through more than one full orbit so that the star has moved past the end of the long axis of its motion where you began, (position C in Figure 1), to produce an overlapping section of the path in the long axis of the motion.

12. Step the star back to the opposite end of the long axis of the ellipse from the overlapping section (to the position marked B in Figure 1).

13. Use the **Hand Tool** to measure the angle between the star and the position half way between the two overlapping parts of the path on the opposite side of the ellipse (the position marked D in Figure 1). Divide this measurement by 2 and record this measurement under the column labeled **Measured Parallax ("")** in Data Table 1, below.

14. Use the formula $d = 1/p$ to determine the distance to the star in parsecs. Multiply this result by 3.26 to convert this distance to light-years. Record these results in the appropriate cells of Data Table 1.

15. Position the hand tool over the star and copy the distance given in the **HUD** into Data Table 1 under the column **Given Distance.**

16. Return to the beginning of Sequence 1 and select another star from the list in Data Table 1.

Data Table 1. Parallax and Distance of a Sample of Nearby Stars

Star	Ecliptic Latitude (°)	Direction of Star's Motion (CW or CCW)	Shape of Star's Motion	Measured Parallax (")	Measured Distance (pc)	Measured Distance (ly)	Given Distance (ly)
Alpha2 Centauri	−42.6						
Bellatrix	−16.8						
T Monocerotis	−16.2						
Procyon	−16.0						
Alpha Sextantis	−11.0						
Spica	−2.0						
Regulus	+0.5						
Capella	+23.0						
Sarin	+47.6						
Delta Cephei	+59.5						
Vega	+62.0						
Polaris	+66.5						

You will note that the closest of these stars, Alpha2 Centauri, is in the southern part of the sky and moves significantly over a year because of parallax.

Question 7. What is the shape and direction of the motion of Alpha2 Centauri in the view?

Question 8. How long does it take for the star to complete this pattern of motion?

It is interesting to compare the distance from the observer of this closest star to the offset distance from the north pole of Earth at which these measurements have been made, to determine if this large offset from the Earth is likely to make a significant difference to the observation of parallax for this close star.

Question 9. Convert the distance to Alpha2 Centauri into AU (1 light-year = 63,240 AU). Expressed as a percentage, what is the ratio between the elevation of the viewing location above the surface of the Earth and the distance to Alpha2 Centauri?

Question 10. Comment on the difficulty in measuring the parallax of T Monocerotis. Can you see any reason why you could measure no parallax motion for this star?

You will notice that Data Table 1 includes a column labeled Ecliptic Latitude. This column of given data indicates the latitude of the star with respect to the ecliptic plane with positive ecliptic latitude indicating that a star is north of the ecliptic plane and a negative ecliptic latitude indicating that a star is south of the ecliptic plane.

Question 11. What relationship, if any, do you see between the rotational direction of the parallactic motion of stars and their ecliptic latitude?

Question 12. What relationship, if any, do you see between the shape of the parallax path and the ecliptic latitude of the star?

Question 13. If a relatively nearby star were to be precisely at the north ecliptic pole, (and ignoring the slow linear drift) would its parallactic path be

a) a straight line back and forth?

b) a perfect circle?

c) an ellipse with the shape of that of the Earth's orbit?

[Optional] If you would like to measure the parallax of other stars, use the following steps.

17. Select the view named **Stellar Parallax from North Pole** from the **Observing Projects > Distance Ladder** folder in the **Favourites** pane.

18. Use the **Find** pane to enter the name of the star whose parallax you would like to measure and when it appears in the **Find** list, click its menu button and select **Centre**.

19. **Zoom In** completely on the star and then position the hand tool slightly to the left or right of the star and click the right mouse button to bring up the **Object Contextual Menu** for the **Celestial Sphere**. Select **Centre** to lock the view onto this point of the sky. The view automatically zooms out to its maximum field of view.

20. Zoom back in on the star and follow the procedure outlined in Sequence 1 above.

C. Spectroscopic Parallax

The next rung of the cosmic distance ladder is the technique of spectroscopic parallax. This method of distance measurement to stars covers the range up to 10 kiloparsecs (about 32,000 light-years). Despite its

> Spectoscopic parallax is discussed in Section 17–8 of Freedman, Geller, and Kaufmann, *Universe*, 9th Ed.

name, this method for determining astronomical distances does not use the measurement of parallax angles. Rather, spectroscopic parallax relies on the concept of standard candles, objects whose luminosity or absolute brightness is known. If we know the absolute brightness of a particular star and compare this to its apparent brightness as seen from the Earth, we can determine the star's distance from Earth using the inverse square law. The method of spectroscopic parallax thus depends on astronomers identifying "standard candles" and determining their absolute brightness.

One such method uses the classification of the spectrum of a star to place it on the Hertzsprung-Russell (H-R) diagram, the widely used summary diagram for stellar astronomy that is discussed in a separate project. The H-R diagram shows the absolute magnitude of a star as a function of its surface temperature. If the star's spectrum can be classified and its temperature and the type of star, (for example, main sequence, giant, or supergiant), established, then it can be placed upon the H-R diagram and its absolute magnitude determined. This absolute magnitude is defined as the magnitude or brightness that this star would have if it were at a standard distance of 10 parsecs or 32.6 light-years from the Sun and is directly related to the star's luminosity or total energy output. With this knowledge of the star's magnitude at a distance of 10 parsecs and a measurement of its apparent brightness, its distance can be determined using the inverse square law. The next sequence of observations uses this method to determine the distances to the stars in Data Table 2 below.

Data Table 2. Spectroscopic Parallax and Distance of a Sample of Nearby Stars

Star	Apparent Magnitude, m	Absolute Magnitude, M	Calculated Distance (pc)	Given Distance (ly)	Given Distance (pc)
Regulus					
Alpha Sextantis					
Polaris					
Delta Cephei					
T Monocerotis					
Zeta Geminorum					

For each of the stars listed in Data Table 2, above, complete the following steps.

> 21. Open the **Favourites** pane and select the appropriately named view from the folder **Observing Projects > Distance Ladder.**
>
> 22. Open the **Info** pane, and under the **Other Data** layer, find the **Apparent magnitude** and **Absolute magnitude** of the star. Record these values in the appropriate columns of Data Table 2 above.

These absolute magnitude values have been established by careful observations of their spectra and other properties to determine their spectral-luminosity classification, in order to place them on an H-R diagram. For example, Regulus is a main sequence B8 star. From its position on the H-R diagram, it has an absolute magnitude of –0.54.

A direct relationship can be derived between the difference in a star's apparent and absolute magnitudes and its distance in parsecs, using the inverse square law and the definition of the magnitude scale for stars.

The total energy emitted by a star, known as its luminosity L (in watts), spreads out into space and the energy per unit area at a distance d meters is the measured brightness of the star, b watts per square meter. At this distance, this total energy is spread uniformly over the surface of a sphere whose area is $4\pi d^2$. Thus, the brightness is

$$b = \frac{L}{4\pi\, d^2}$$

This is known as the **inverse square law** governing the change of radiation intensity as a function of distance away from the source.

Two stars of brightness b_1 and b_2, at distances d_1 and d_2, respectively, whose luminosities, or total energy output, L_1 and L_2, are related by

$$\frac{b_1}{b_2} = \frac{L_1}{L_2} \left[\frac{d_2}{d_1}\right]^2$$

This ratio will hold in any units of distance, whether in meters or parsecs.

The difference between the magnitudes of two stars is a measure of the ratio of their brightnesses, such that a difference of $m_2 - m_1 = 1$ magnitude corresponds to a ratio of $b_1/b_2 = 2.512$. This odd value is chosen so that a 5-magnitude difference is exactly 100 [i.e., $(2.512)^5 = +100$]. Furthermore, the magnitude scale is inverted, such that the larger the magnitude, the smaller the ratio. Brighter stars have lower magnitudes!

Thus, the equation relating two stars of magnitude m_1 and m_2, whose brightnesses are b_1 and b_2, respectively, is

$$m_2 - m_1 = 2.512 \times \log\left[\frac{b_1}{b_2}\right]$$

If now we want to relate the brightness of the same star that is measured from two distances, the distance, d, from which the star has a magnitude of m, and the distance of 10 parsecs, from which the star would have absolute magnitude M, we can use this relationship and the inverse square law. From the latter, the ratio of brightnesses, b from distance d, and B from a distance of 10 parsecs for the same star in which $L_1 = L_2$, is

$$\frac{b}{B} = \left[\frac{10}{d}\right]^2$$

In magnitudes, this can be represented as

$$m - M = 2.512 \times \log\left[\frac{B}{b}\right]$$

$$= 2.512 \times \log\left[\frac{d}{10}\right]^2$$

$$= 2 \times 2.512 \times \log\left[\frac{d}{10}\right]$$

The constant 2.512 is often simplified to 2.5 in calculations. Using this approximation, and recognizing that the logarithm of a ratio is the difference of the logarithms of the components of the ratio, this equation becomes

$$m - M = 5\left[\log d - \log 10\right]$$

Thus, since $\log 10 = 1$,

$$m - M = 5\log d - 5 \qquad\qquad \text{(Equation 1)}$$

Thus, d, in parsecs, can be derived from knowledge of m and M. This difference, $m - M$, between a star's apparent magnitude as seen from Earth, m, and its absolute magnitude, M, is called the **distance modulus** of the star.

The apparent magnitude of the star is measured through observation. By rearranging the distance modulus formula, the distance to the star (in parsecs) can be calculated as follows:

$$d = 10^{\left[\frac{m - M + 5}{5}\right]} \qquad\qquad \text{(Equation 2)}$$

23. Use the formula in Equation 2 above to calculate the distances to the stars listed in Data Table 2. Take care with the sign of the magnitude values. Compare your results with those you obtained in Data Table 2 for those stars that are in both tables.

Question 14. Did you encounter any difficulty determining the distance to T Monocerotis using spectroscopic parallax?

Question 15. Which method of determining stellar distances has the greater range, stellar parallax or spectroscopic parallax?

The two methods for determining the distance to a star can be used to verify each other. For example, we can calculate the absolute magnitude of Regulus from its apparent magnitude using the inverse-square law and the distance, d, in parsecs, to Regulus as determined by stellar parallax. To do so, simply rearrange Equation 1 to solve for M as shown in Equation 3, below:

$$M = m + 5 - 5\log(d) \qquad\qquad \text{(Equation 3)}$$

Question 16. a) Using the value for the distance (in parsecs) that you obtained by measuring the stellar parallax of Regulus in Data Table 1, calculate this star's absolute magnitude using Equation 3 above, and the given value for the apparent magnitude of Regulus that you recorded in Data Table 2. What is the absolute magnitude of Regulus as determined from its apparent magnitude and distance as measured by its parallax? b) How does your result compare to the absolute magnitude of Regulus as determined from its spectral characteristics?

Question 17. You can verify the spectroscopic parallax of a star against its measured stellar parallax in the same way as in the previous question by using the distance to the star determined from its spectroscopic parallax to predict its stellar parallax. Using the distance (in parsecs) determined by its spectroscopic parallax, calculate the stellar parallax for the following stars: a) delta Cephei b) T Monocerotis.

Question 18. How does the predicted stellar parallax of delta Cephei compare to the parallax that you measured?

Question 19. a) Would the *Hipparcos* satellite be able to measure the stellar parallax of T Monocerotis? b) How do you think the distance to T Monocerotis that is given in *Starry Night Enthusiast*™ was determined?

D. Population I Cepheid Variables

Cepheid variable stars are stars that have evolved away from the main sequence, and occupy the so-called instability strip on the H-R diagram. Cepheids vary regularly in brightness in a specific and identifiable pattern. They received their name from the prototype for this class of stars, δ (delta) Cephei.

There are two features of Cepheid variable stars that make them extremely useful for measuring distances in the universe. First of all, the period over which the brightness of a Cepheid varies is related directly to its average intrinsic luminosity, or absolute magnitude. Secondly, Cepheids are intrinsically very bright, with luminosities ranging up to over 10,000 times the luminosity of the Sun. Consequently, they can be seen from distances of millions of parsecs and their apparent magnitudes measured. Combination of these measured values with their absolute magnitude determined from the period-luminosity relation using the equations above provides distances over much larger distances than the previous methods, thereby extending the range of the distance ladder significantly.

Data Table 3 contains a sample of Cepheid variable stars and includes data on each star's period of variability. For each star in this list, follow the steps below.

24. Open the **Favourites** pane and select the appropriately named view from the **Observing Projects > Distance Ladder** folder.

25. Open the **Info** pane and copy the values for the star's **Absolute magnitude** into Data Table 3 below.

Data Table 3. Period and Absolute Magnitudes of a Sample of Cepheid Variable Stars

Star	Period (days)	Absolute Magnitude
Zeta Geminorum	10.15	
TT Aquilae	13.75	
Eta Aquilae	7.18	
T Vulpeculae	4.44	
DT Cygni	2.50	

26. Plot the data in Data Table 3 onto the graph in Figure 1 below.

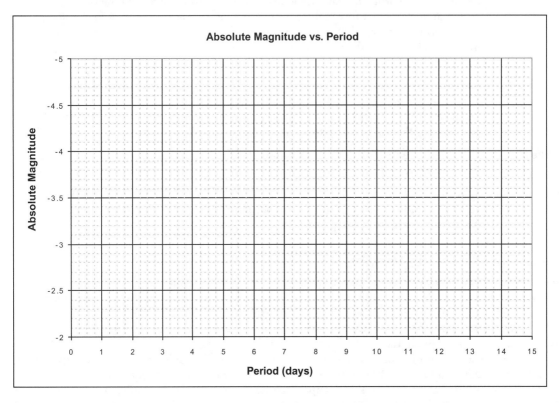

Figure 1. Graph of Absolute Magnitude versus Period for a Sample of Cepheid Variables

Question 20. What type of relationship does your graph suggest between the period of variability and the absolute magnitude of Cepheid variable stars?

27. Attempt to draw a straight line through your data points.
28. [Optional] Use a spreadsheet application such as Excel that includes charting capability to create this graph. The following procedure can be used for plotting this graph in Excel. Copy the data from Data Table 3 into the cells of an Excel worksheet. Highlight the **Period** and **Absolute Magnitude** values in the spreadsheet and select **Insert > Chart...** In step 1 of the Chart Wizard, select the **XY (scatter)** chart. Click the **Next** button. In step 2 of the Chart Wizard, select **Series in... Columns** and click **Next**. In Chart Wizard – Step 3, click the **Titles** tab and enter appropriate titles for the graph and the axes. Click the **Legend** tab and clear the checkbox beside **Show Legend**. Click the **Gridlines** tab and check all of the checkboxes; then click the **Next** button. In Chart Wizard – Step 4, click the **Finish** button. In the chart that appears, click on the vertical axis showing **Absolute Magnitude** to highlight it. Then select **Format > Selected axis...** from the menu. In the dialog box that pops up, click the **Scale** tab and set the **Minimum** value to –5, **Maximum** value to –2, **Major Unit** to 1, **Minor Unit** to 0.2 and then check the box beside the **Values in reverse** order option. Then click **OK**. Click the horizontal axis of the chart to highlight it and select **Format > Selected axis...** from the menu. In the dialog box that pops up, click the **Scale** tab and set the **Minimum** value to 0, **Maximum** value to 15, **Major Unit** to 1, **Minor Unit** to 0.25. Then click the Patterns tab and set the **Major tick mark type** option to **Inside** and set the **Tick mark labels** to the **High** option. Then click **OK**. Finally, select **Chart > Add Trendline...** from the menu. In the **Add Trendline** dialog box, select the **Linear Trend/Regression** type and in the **Options** dialog box, select **Display equation on chart**. Click the **OK** button. Print out this graph of your data and the trendline showing the relationship between a Cepheid variable star's period and its absolute magnitude, or luminosity.

If the relationship that you plotted in Figure 1 is valid, you should be able to find the distance to another Cepheid variable simply by observing the period of its variability. With a measurement of this period you can find its absolute magnitude or luminosity and by comparing this absolute magnitude to its apparent magnitude, determine its distance from the inverse-square law. This technique is similar to spectroscopic parallax but depends on the period-luminosity relationship rather than its spectral characteristics and the H-R diagram for determining the absolute magnitude of the star. Moreover, the distance you determine for a Cepheid variable from this period-luminosity relationship should agree with its distance as determined by other methods, particularly the method of stellar parallax. You have already determined the distance for Delta Cephei, the star for which all the other stars in the class of Cepheid variables is named, by measuring its stellar parallax and calculating its spectroscopic parallax. Follow the next steps to determine the distance to this star from the period-luminosity relationship.

29. The period of the Cepheid variable star Delta Cephei is 5.37 days.

30. Locate the point for a period of 5.37 days on the horizontal axis of your graph and mark where it intersects the trendline that you drew through the graph in Figure 1. Read the corresponding absolute magnitude for this star as determined from your graph. Use this absolute magnitude value and the apparent magnitude of Delta Cephei ($m = 4.06$) in Equation 2 above to determine its distance.

Question 21. a) What is the absolute magnitude of Delta Cephei based upon the period-luminosity graph? b) Does this agree with the absolute magnitude of this star in Data Table 2, as determined from its spectroscopic parallax?

Question 22. a) What is the distance to Delta Cephei based on its period of variability? b) How does this compare with the distance you obtained for this star by measuring its parallactic shift?

The astronomer Henrietta Leavitt used observations of a much larger sample of Cepheid variables when she established this period-luminosity relationship in 1912. Since her work, two different types of Cepheid variable stars have been identified: Population I and Population II Cepheids. Each shows a specific period-luminosity relationship. Thus, in addition to identifying a star as a Cepheid, astronomers must also determine to which population the star belongs before a definitive luminosity can be assigned to it.

Another class of variable stars, the population II RR Lyrae stars, also show a period-luminosity relationship. RR Lyrae stars are generally not as intrinsically luminous as Population I Cepheid variables, and have a very narrow but well-defined range of luminosities. While they do not provide as great a range of distance measurement as Cepheids, they are much more common and have therefore been important in the verification of other rungs in the cosmic distance ladder. Also, in 1920, Harlow Shapley used the period-luminosity relationship of RR Lyrae stars in globular clusters to determine the distances to these star groups in our Milky Way Galaxy. From the distances and distribution of these globular clusters, he was able to deduce the size and extent of the Milky Way Galaxy.

As mentioned above, the other reason that Cepheid variables are so important in the distance ladder is that they are intrinsically highly luminous stars, visible at vast distances from Earth. Cepheid variable stars can even be resolved in other galaxies. In 1923, Edwin Hubble used the 100-inch telescope at Mount Wilson to observe Cepheid variables in what was then called the Andromeda nebula, an object considered to be in our own Galaxy. From the period-luminosity relationship, Hubble determined that this "nebula" was in fact an independent star system approximately 2 million light-years distant from the Sun. Since Hubble's time, Population I Cepheid variables have extended the distance ladder to a range of up to 30-megaparsecs (Mpc), which is equivalent to almost 100 million light-years. Thus, Cepheid variable stars provide a crucial foundation for the next rung in the distance ladder.

E. The Hubble Law

There are several other distance determination methods that astronomers use that overlap and otherwise corroborate those we have examined so far in this project. These include the use of masers, the Tully-Fisher relation and the use of Type IA supernovae as standard candles. Notwithstanding these intermediate methods for distance measurement to neighboring galaxies, it is the Cepheid variable method that provides the direct link of distances to galaxies and the link to the final rung in the distance ladder: the Hubble law.

A few years after uncovering the true nature and distance to the Andromeda Galaxy, Edwin Hubble discovered a relationship between the distance to a galaxy and the speed at which it is receding, as measured by the redshift of the lines in its spectrum. This relationship, called the Hubble law, has allowed astronomers to estimate distances to the limits of the observable universe. Furthermore, this relationship has profoundly altered our knowledge of the evolution of the universe.

A significant feature of Hubble's results is that, apart from a few nearby galaxies such as the Andromeda Galaxy and members of the Local Group, which appear to be gravitationally bound together, most other galaxies in the universe appear to be receding from each other. Cosmologists, supported by Einstein's general theory of relativity, have been able to conclude that this redshift is a consequence of the expansion of space itself. By extrapolating this expansion into the past, cosmologists have suggested that the universe was born 13–15 billion years ago from the explosion of an infinitesimal point of infinite density. Supporting evidence for this Big Bang cosmology has come from the study of the residual radiation of this explosion, the cosmic microwave background radiation.

The Hubble law states that the distance to a galaxy, d, is proportional to its velocity of recession, v, and is simply expressed in the formula

$$d = \frac{v}{H_0}$$
(Equation 4)

where H_0 is a constant called the Hubble constant and is expressed in units of kilometers per second per megaparsec. The determination of the Hubble constant has been an on-going process, with early measurements from different techniques providing widely varying values. Recent work has now determined its value with reasonable accuracy to be 70 km/s/Mpc.

Astronomers continue to use other rungs of the distance ladder, particularly Type IA supernovae and the Tully-Fisher relation, to calibrate this important cosmological constant.

Before stepping briefly on this final rung of the distance ladder, it is perhaps instructive to see how far our distance measuring techniques have taken us.

31. Select **Favourites > Observing Projects > Distance Ladder > Parallax to Hubble Law.**

The view shows the Sun and the planets.

> **Question 23.** Which rung of the distance ladder is appropriate for determining distances in the solar system?

32. **Increase current elevation** to about **32 ly from Sun.** Use the **Location Scroller** to look around the view. (Hold the **Shift** key down while clicking and moving the mouse. The 3-dimensional structure of this region of space will be apparent as you do this.) This is equivalent to moving around on the surface of a sphere centered on the Sun whose radius is 32 ly.

33. While using the **Location Scroller,** you will note that stars appear to move by different amounts. You can use the **HUD** to determine the distances of these stars from the observer. You will note that those closest to the observer appear to move the most as you move around the **Location Scroller** sphere.

> **Question 24.** What distance determination techniques would be suitable for the majority of objects in this view?

34. **Increase current elevation** to about **3200 ly from Sun.** Use the **Location Scroller** to look around the view.

The local neighborhood of stars around the Sun and the surrounding disk of the Milky Way are visible in the view.

> **Question 25.** Which distance determination technique is most suitable for objects that are at the distance of the current viewing location from the Sun?

35. **Increase current elevation** to about **0.326 Mly from Sun.** Use the **Location Scroller** to look around the view.

The Milky Way and the nearby Andromeda Galaxy are labeled in the view. All of the other points of light in the view are other galaxies. Some of these galaxies, such as the Andromeda Galaxy, are near enough to the Earth that astronomers can use the very luminous Cepheid variables and the period-luminosity relation to determine the distances to these galaxies. Beyond the reach of Cepheid variables, other, even more luminous, standard candles such as Type IA supernovae, extend our distance measuring capability to even more distant galaxies, out to about 3.2 billion light-years.

36. **Increase current elevation** to about **326 Mly from Sun** and look around the view with the **Location Scroller.**

At this distance, you are seeing the large-scale structure of our universe, with walls of galaxies surrounding huge voids of space containing few galaxies. Other corroborative distance measuring techniques can be used at this range but we are now firmly within the realm of the Hubble law.

37. **Increase current elevation** to about **3260 Mly from Sun** and look around the view with the **Location Scroller.**

At this distance from the Sun, the database of galaxies in *Starry Night Enthusiast*™ has reached its limit. The boundary you see in the view is artificial. The empty space surrounding the cube of galaxies in the view is itself filled with billions of other galaxies out to the limits of our observation at about 12 billion light-years from the Earth. You will note the galaxy labeled NGC 2484 near the boundary of the *Starry Night Enthusiast*™ database limit. This galaxy is receding from us and has a measured redshift, z, of 0.042836. Use this to answer the following questions.

> **Question 26.** Using the formula $v = z \times c$, where z is the redshift, c is the speed of light, what is the recessional velocity, v, of NGC 2484?

38. **Zoom In** to a field of view about **25° wide** and use the **Location Scroller** to adjust the view so that you can identify NGC 2484 with the **Hand Tool.** Then, with the **HUD** identifying the object at the cursor as NGC 2484, open the **Object Contextual Menu** for this galaxy and select **Magnify.** Use the **Location Scroller** to observe the size and shape of this galaxy.

> **Question 27.** Assuming a value for the Hubble constant, H_0, of 75 km/s/Mpc, what is the distance to this galaxy from the Sun a) in Megaparsecs (Mpc)? b) in light-years?

39. Select **Favourites > Observing Projects > Distance Ladder > NGC 2484.**

The view shows the galaxy NGC 2484 in the direction of the Sun and the Milky Way from a viewing location in space almost 3.3 billion light-years from the Sun. In the right upper corner of the view, *Starry Night Enthusiast*™ displays the distance of the current viewing location from the Sun and from NGC 2484.

Question 28. In this view, which is farther away, the Sun in the Milky Way or NGC 2484?

Question 29. From the distances displayed in the upper right corner of the view, what is the distance between the Sun and NGC 2484 as given in *Starry Night Enthusiast*™?

Question 30. What value does *Starry Night Enthusiast*™ use for the Hubble constant?

F. Conclusions

In this project, you have had the opportunity to review the techniques that astronomers employ to measure distances to objects in the cosmos. You had the opportunity to employ the basic principles of these techniques in software simulations. Finally, you learned about the distance ladder and how its various rungs not only extend the range of measurable distances but also verify and calibrate other rungs when they overlap in applicable ranges.

The Milky Way

21

One feature of the night sky that must have attracted the attention of the very first "astronomers" is the ribbon of light that encircles it, the Milky Way. Sadly, the light pollution of today's modern world robs many of us of the opportunity to see this beautiful, subtle light that originates from within our home Galaxy in the cosmos.

Galileo made the first telescopic observations of the Milky Way in 1609. His telescope resolved the diffuse band of light into countless individual stars. Later telescopic observers, such as Charles Messier, began to discover other objects in the sky that appeared fuzzy in the eyepiece and could not be resolved into stars. Little was known of these mysterious objects except that some were eventually resolved into clusters of stars, while the rest remained nebulous. In the past century, observations showed that some of the nebulous objects were clouds of gas and dust within our own Galaxy, while others were themselves distant galaxies.

> The Milky Way is discussed in Chapter 23 of Freedman, Geller, and Kaufmann, *Universe*, 9th Ed.

Modern observations of these star clusters, nebulae, and galaxies have helped astronomers to:

1) establish accurate estimates of the physical size of the Galaxy,

2) develop an understanding of its physical and dynamic structure,

3) determine that our Milky Way Galaxy is but one of billions of such galaxies in the universe, and

4) construct theories about the origin and structure of the universe, based on the observed filamentary distribution and relative motion of the billions of galaxies within it.

In this project, you will use *Starry Night Enthusiast*™ to make observations of the Milky Way and some of these nebulous telescopic objects.

A. The Milky Way and the Broad Structure of the Galaxy

The term "Milky Way" refers, in its strictest sense, to the appearance of the Galaxy as seen from Earth, specifically to the diffuse ribbon of luminosity that winds its way among the constellations in our sky.

> 1. Launch *Starry Night Enthusiast*™.
> 2. Select **Favourites > Observing Projects > Milky Way > Cornwall**.

The view is of the southern sky from Cornwall, Canada. The time is midnight on the date of the summer solstice. The teapot and fish hook asterisms are visible near the southern horizon and the band of the Milky Way sweeps diagonally upward to the left.

3. Select **Options > Stars > Milky Way...** from the menu and, in the dialog window that pops up, adjust the brightness of the Milky Way to its maximum in the view, using the slide control.

Were you to go outside at the time and place of this simulation and look south at the teapot asterism, you would see the Milky Way appearing to rise like steam from the teapot's spout, then sweep upward across the sky toward the summer triangle near the zenith and finally arc back down through the "W" of Cassiopeia in the northeast before reaching the horizon again.

4. Use the **Hand Tool** to drag the sky in a manner that allows you to follow the path of the Milky Way as described in the previous paragraph.

You can see that the Milky Way forms a band of varying width across the sky from horizon to horizon. In fact, the Milky Way completely encircles the sky as seen from Earth.

5. Open the **Options** pane and click the checkbox next to the **Local Horizon...** option to remove the horizon from the view.
6. Use the **Hand Tool** to drag the view and follow the complete path of the Milky Way around the sky.

Question 1. Does the Milky Way encircle the sky as a continuous band, or is it discontinuous, disappearing and reappearing at intervals?

7. Select **View > Constellations > Astronomical.**
8. Use the **Hand Tool** to drag the view and again follow the complete path of the Milky Way around the sky. As you do so, observe the size and brightness of the Milky Way.

Question 2. Toward which constellations does the Milky Way appear broadest?

Question 3. In the direction of which constellations does the Milky Way appear narrowest?

Speculate on your observations. The fact that the Milky Way encircles the sky as seen from Earth suggests that the solar system is somewhere inside the Milky Way Galaxy. If this were not the case, then the Milky Way would appear discontinuous and would be confined to a particular direction in the sky. The fact that the Milky Way appears as a relatively narrow band suggests that the Milky Way Galaxy has a flat structure. From observations of other galaxies, we know that such a flat, disk-like structure is characteristic of spiral and barred spiral galaxies. Consequently, we can deduce that the Milky Way Galaxy is a type of spiral galaxy. If it were elliptical or irregular, stars would be dispersed across the sky instead of being concentrated in such a narrow band. If this deduction is correct, then the Milky Way Galaxy must also have a central bulge. As seen from our position on Earth inside the Galaxy, this bulge would likely reveal itself as a broader region in the Milky Way.

Question 4. Using the logic of the previous paragraph and your answer to Question 2, toward which constellation(s) must an observer on Earth look to see the center of the Milky Way Galaxy?

Question 5. Toward which constellation would an observer on Earth need to look in order to look directly out of the Galaxy, away from the center along the plane of its disk?

The Milky Way | **241**

9. Select **File > Revert** and then press the **K** key on the keyboard to remove the asterism outlines and labels.

10. Set the **Gaze** to the South.

11. Change the **Time flow rate** to **300×** and follow the position of the Milky Way along the horizon as time advances. Daylight has been hidden to allow you to follow the variable angle that the Milky Way makes to the horizon at different times.

The reason for the variable angle and rising point of the Milky Way is that the plane of the Galaxy is inclined to the equatorial plane of the Earth.

12. Select **Favourites > Observing Projects > Milky Way > Galactic Plane**.

The lines in the view represent three significant planes. The red line is the celestial equator and represents the projection of the plane of Earth's equator onto the celestial sphere. The green line is the ecliptic, the plane of Earth's orbit around the Sun. The blue line is the galactic equator, the plane of the disk of the Milky Way Galaxy. The angles between these various planes can be obtained by measuring the angles in the sky between the poles of these planes.

13. Open the **Find** pane and type **Alkaid** in the search box and press **Enter**. This centers the star Alkaid in the view.

Three points are marked in the sky, the north celestial pole, the north ecliptic pole and the north galactic pole.

14. You can estimate the angle at which the Earth's rotation axis is inclined to the plane of the Galaxy by using the **Angular Separation Tool** to measure the angle between the star Polaris (near the north celestial pole) and the north galactic pole (or a point very close to it).

15. You can estimate the angle at which the ecliptic is inclined to the galactic plane by measuring the angular separation between the star Alkaid, first to the north galactic pole then to the north ecliptic pole, and add the two values.

Question 6. What is the approximate angular separation between the north celestial pole and the north galactic pole?

Question 7. At approximately what angle does the galactic plane intersect the celestial equator?

Question 8. What is the approximate angular separation between the north ecliptic pole and the north galactic pole?

Question 9. At approximately what angle does the ecliptic intersect the galactic plane?

B. Details of the Naked-Eye Structure of the Milky Way

You probably noticed some interesting dark encroachments into the band of the Milky Way in the observations you made in the previous section.

16. Select **Favourites > Observing Projects > Milky Way > Great Rift**.

In this view to the south, again from Cornwall, Canada, at midnight on June 21, the Milky Way stretches diagonally across the sky. Beginning at the star Deneb in Cygnus and extending to the star Nu Ophiuchi (ν Oph) in the lower right, the Milky Way is split by a dark intrusion known as the Great Rift. The northern part of this structure is known as the Cygnus Rift and also as the Northern Coalsack for a reason that you will discover shortly.

The Great Rift is representative of a broad class of objects within the Galaxy called **dark nebulae.** Dark

> Dark nebulae are discussed in Section 18–2 of Freedman, Geller, and Kaufmann, *Universe,* 9th Ed.

nebulae reveal themselves by obscuring the light of a radiant background such as the rich star fields of the Milky Way. As you can see, the Great Rift creates the illusion that the Milky Way splits into two branches.

Dark nebulae range in size from relatively small, almost spherical Bok globules to immense clouds of gas and dust such as the Great Rift. Modern estimates suggest that the mass of the gas and dust contained in the Great Rift is equivalent to 1 million Suns!

17. Measure the angular separation between Deneb and nu Ophiuchi (ν Oph). Use this measurement to estimate the approximate angular extent of the Great Rift.

Question 10. What is the approximate angular extent of the Great Rift in our sky?

For southern hemisphere observers, an equally impressive dark nebula is the Coalsack, near to the Southern Cross. Like its northern namesake, the Coalsack is easily seen and appreciated with the naked eye, standing in such stark contrast to the rich star fields of the southern Milky Way that it gives the illusion of being darker than the rest of the sky.

18. Select **Favourites > Observing Projects > Milky Way > Coalsack.**

The Coalsack, named for its darkness and its shape, is visible below the Southern Cross asterism.

19. To identify the Southern Cross asterism, Select **View > Constellations > Asterisms** and **View > Constellations > Labels** from the main menu. Press the **K** key to toggle the asterism and label display off again.

20. To get a better view of the Coalsack, center it in the view and then **Zoom In** to a field of view about 9° wide.

21. Open the **Info** pane and read the information under the **Description** layer.

Question 11. What is the approximate area of sky that is obscured by the Coalsack dark nebula?

Question 12. By way of comparison, the Full Moon, as seen from Earth has an angular radius of about 15' (0.25°). How much larger in area in the sky is the Coalsack than the Full Moon?

> Giant molecular clouds are discussed in Section 18–7 of Freedman, Geller, and Kaufmann, *Universe,* 9th Ed.

Dark nebulae are visible evidence for the existence of interstellar matter in the Galaxy. Roughly half of this interstellar matter is in the form of immense, gravitationally bound structures known as **giant molecular clouds.** These vast clouds typically contain a mass of more than 10,000 Suns, primarily in the form of hydrogen and helium gas, plus a small fraction of heavier elements and dust. The dust consists of tiny, solid particles of graphite or silicates.

The interstellar medium is extremely rarefied, but despite its low density, its pervasiveness in the vast dimensions of the Galaxy has lead astronomers to estimate that the matter of the interstellar medium amounts to approximately 10 percent of the total mass of the Milky Way Galaxy.

C. HI and HII Regions

The dark nebulae are often called **HI regions** (pronounced "H one"). HI is the symbol used to describe un-ionized hydrogen. These nebulae are cool (about 100 K) and the hydrogen they contain is therefore neutral rather than ionized. Astronomers have learned that dark nebulae are often the birthplaces of stars. If some of these newborn stars are hot, massive O and B stars, their ultraviolet light ionizes the hydrogen in the part of the HI region surrounding them to form an **HII region** (pronounced "H two," this symbol describes ionized hydrogen). Because of the energy absorbed from the ultraviolet light, HII regions are hot, typically about 10,000 K. Subsequent recombination of the electrons and ions causes HII regions to emit light at visible wavelengths, predominantly in the strong Balmer-α spectral line in the red region of the spectrum. Because of this emission, HII regions are also called **emission nebulae.**

> HI and HII regions are discussed in Section 18–7 and emission and reflection nebluae in Section 18–2 of Freedman, Geller, and Kaufmann, *Universe*, 9th Ed.

The HI regions surrounding HII regions often appear as silhouettes. They can also become visible by reflecting the light of nearby hot, young stars. Since the dust grains within these cold HI clouds preferentially scatter light of shorter wavelength, these **reflection nebulae** shine with a blue color.

22. Select the **Favourites > Observing Projects > Milky Way > M8, M20 and M21.**

There are two different HII regions, M8 and M20, within this 5° field of view, in addition to the open cluster of stars designated M21.

23. Use the **Object Contextual Menu** over M8 (Lagoon Nebula) and select **Magnify** and **Show Info.** In the Info pane, expand the **Description** layer and read about this HII region.
24. **Zoom Out** to a field of view about 5° wide and repeat the previous instruction for the object labeled M20 (Trifid Nebula).

You will notice that both of these objects contain rich open clusters of newborn stars.

25. **Zoom Out** to a field of view about 5° wide and repeat the previous instruction for the object labeled M21.

Question 13. Approximately how old is the star cluster that forms M21?

Question 14. Approximately how many stars are readily visible in M21 when viewed through a small telescope?

26. Select **Favourites > Observing Projects > Milky Way > M42.**

M42, the great Orion Nebula, is a region of very young stars within a giant molecular cloud whose total mass is estimated to be approximately 500,000 times that of the Sun. About 1600 light-years from Earth, M42 lies on the edge of its parent giant molecular cloud, which in turn is but one of a vast system of such clouds found in this part of the Galaxy. Near the central portion of M42 lies the Trapezium, named for the four brightest components of the system of recently formed O- and B-type stars that are providing the radiation that lights up the surrounding nebulae. Parts of the parent molecular cloud near M42 are prolific stellar nurseries, where new clusters of stars are actively forming.

27. In this magnified view of M42, use the **Hand Tool** to identify a member star of the Trapezium and, using the **Object Contextual Menu, Centre** it in the view. Then **Zoom In** to a field of view about 10' wide to view the four stars that give the Trapezium its name.

28. Select **File > Revert.**

29. Using a member star of the Trapezium as a starting point, use the **Angular Measurement Tool** to estimate the angular radius of M42.

> **Question 15.** What is the approximate average angular radius of M42?
>
> **Question 16.** At its distance of 1600 light-years from Earth, what is the approximate physical radius of M42 in light-years? [*Hint:* Use the small angle formula.]

HII regions make splendid targets for small and large telescopes alike, forming large, bright canvasses against which the marvelously complex morphology of the surrounding, colder, dark HI clouds is frequently silhouetted. When viewed (and particularly when photographed) through a telescope, the combination of newborn stars, dark nebulae, reflection nebulae and HII regions make the stellar nurseries of the Galaxy provocatively beautiful.

30. Select the **Favourites > Observing Projects > Milky Way > Horsehead.**

31. **Zoom In** on the field to see how this silhouette of an HI region against a bright HII region got its name.

32. Open the **Info** pane and read the **Description** of this object.

This region of the sky is a perfect example of the effect of dense dust clouds obscuring more distant stars. The region of the sky to the left of the bright emission nebula appears to contain far less stars than does the region of the nebula. This is because this part of the nebula is a dense HI region containing large quantities of absorbing dust, which has obscured the background stars in this region of the sky.

D. Evolution of HII Regions

Stellar winds from newborn stars at the core of an HII region create shock waves that can trigger new regions of star formation within the surrounding dark nebula.

33. Select **Favourites > Observing Projects > Milky Way > M42.**

In the magnified image of M42, you can see such a shock wave, which appears as a slightly brighter edge in the boundary between the blue reflection nebula and red HII region in the lower left part of the image.

The lifetime of a typical HI region is only several million years because its gas and dust are consumed over time in the formation of new stars and HII regions. The gas in an HII region is hot and is not gravitationally bound and so expands outward. Eventually, very little gas and dust remains, leaving only a cluster of newborn stars in its wake.

34. Select **Favourites > Observing Projects > Milky Way > M16.**

M16, popularly called the Eagle Nebula because of the shape of the dark nebula silhouetted against it, is an HII region approximately 7000 light-years from Earth toward the inner part of the Galaxy. The members of the embedded cluster of stars are estimated to be less than 1 million years old.

35. Open the **Info** pane and read the **Description** of this HII region.

36. Open the **Sky Guide** pane and click on the underlined link labeled **Enter SkyGuide**. Then navigate through the following links: **Night Sky Tours > Twenty great fuzzies > The Eagle nebula**. Read the information on this object and click on the link **pillars of creation** to see the Hubble Space telescope image of this object and read a further description of it.

In this Hubble image, protostars can be seen as they emerge from their enveloping cocoons of gas and dust.

Question 17. How long are the gaseous towers in this Hubble image of the Eagle Nebula?

E. Open Clusters

> Open clusters are discussed in Section 18–6 of Freedman, Geller, and Kaufmann, *Universe*, 9th Ed.

Open clusters are the progeny of HI regions. On the whole, the member stars of an open cluster are gravitationally bound to one another. Over time, however, some members of the cluster, perhaps nudged by other stars in the cluster, escape the gravitational hold of their siblings. This reduces the overall mass of the cluster as a whole, thereby decreasing its gravitational hold on the members that remain. Thus, an open cluster gradually "evaporates" over a time of several million to several billion years as its member stars disperse through the Galaxy. Nevertheless, virtually all of the stars in the disk of the Galaxy were once members of such "families" of stars.

Several open clusters are visible to the naked eye and have been known since antiquity. These include the Pleiades (the Seven Sisters) and the Hyades, both in Taurus, and Praesepe (the Beehive) in Cancer.

37. Select **Favourites > Observing Projects > Milky Way > M45**.

In this telescopic view of the Pleiades, the brighter members of the cluster gleam brilliantly, their light reflected in residual strands of the gas and dust from which they were born. As with all reflection nebulae, the reflected light is predominantly blue.

38. Select **Favourites > Observing Projects > Milky Way > Open Clusters**.

Three open clusters, all in the constellation Auriga, are visible in this field of view, M36, M37, and M38. All three clusters are approximately the same distance from the Earth, about 4000 light-years. This fact suggests that they might all have been born within the same HI region.

39. Use the **Object Contextual Menu** and select **Magnify** and **Show Info** for each of these objects to see a close-up image of these star clusters and read their **Description** in the **Info** pane.

Question 18. What is the estimated age of the open star cluster M36?

F. Globular Clusters

> Globular clusters are described in Section 23–1 of Freedman, Geller, and Kaufmann, *Universe*, 9th Ed.

The Galaxy contains another type of star cluster, the **globular cluster,** which gets its name from its strongly condensed, distinctly globular shape. Unlike open clusters, which are concentrated in the spiral arms of the Galaxy's disk, globular clusters are distributed in a vast and roughly spherical halo around the Galaxy's central bulge.

Globular clusters are much older and much richer than their open counterparts, possessing tens of thousands to millions of member stars. The spectra of the stars in globular clusters show them to be metal-poor (to an astronomer, any element heavier than helium is a metal!), their chemical composition being not far removed from the abundance of elements produced in the Big Bang. In contrast, the stars of open clusters, and virtually all of the other stars in the galactic disk, have metal-rich spectra. Another feature of globular clusters is the virtual absence of dust and gas within the cluster, in distinct contrast to open clusters.

From their age and distribution within the Galaxy, globular clusters are thought to have formed from the central condensation of matter in the protogalaxy that became the Milky Way. Globular clusters are spectacular sights when seen through a telescope.

> 40. Select **Favourites > Observing Projects > Milky Way > Omega Centauri.**

The view shows a telescopic image of the finest globular star cluster in the sky, Omega Centauri, catalogued as NGC 5139. This cluster of over 1 million stars is approximately 16,000 light-years from Earth in the direction of the constellation Centaurus. As you can see, globular clusters are aptly named.

> 41. Open the **Info** pane and read the **Description** of Omega Centauri. Also note the **Angular size** of this object in the sky, as given in the **Other Data** layer of the **Info** pane.

Question 19. Using the angular size of this object as provided in the **Info** pane as its angular diameter, what is the approximate angular radius of the globular cluster Omega Centauri?

Question 20. Given that the distance to this object is 16,000 light-years, what is the physical radius of Omega Centauri in light-years? [*Hint:* Use the small-angle formula.]

Question 21. If we assume that Omega Centauri is spherical and that it contains 1 million stars, what is the average density of the cluster in units of stars per cubic light-year? [*Hint:* The volume of a sphere of radius R is $4\pi R^3/3$.]

Question 22. Approximately what is the average distance between stars in this globular cluster? [*Hint:* Take the reciprocal of the answer to the previous question, which is the volume of space occupied by a star on average. If you assume that each star occupies a cubic volume of dimension d, its volume will be d^3. Thus, take the cube root of this volume per star to derive d, the spacing between stars.]

G. Planetary Nebulae and Supernova Remnants

Planetary nebulae are discussed in Section 20–3 of Freedman, Geller, and Kaufmann, *Universe*, 9th Ed.

After stars are born in the stellar nurseries of HI regions, their subsequent lifetime on the main sequence, as well as their destinies once the fuel in their cores is exhausted, depends on their initial mass. Low-mass stars, those whose initial masses on the main sequence were less than eight solar masses, die by gently ejecting their outer layers into the interstellar medium to produce so-called **planetary nebulae.** The ejection is caused by bursts of nuclear energy in shells of matter near the boundary of the star's core.

The material of these ejected shells consists of hydrogen and helium, as well as atoms of heavier elements such as carbon and oxygen that were produced in various phases of the star's evolution after it left the main sequence. These atoms absorb radiation emitted from the dying parent star and subsequently re-emit it in the characteristic wavelengths for that element, producing the glowing shell of gas that we see as a planetary nebula.

The ejection of the outer part of the star leaves behind the very hot central core, which is now the central "star" of the planetary nebula. By this time, all nuclear burning within the core has ceased, and it remains behind as a small, but very hot, white dwarf star. The concentric shells of gas forming the planetary nebula gradually dim as they disperse into the interstellar medium, the whole visible structure lasting about 50,000 years.

42. Select **Favourites > Observing Projects > Milky Way > M57.**

43. Open the **Info** pane and expand the **Description** layer to learn more about M57.

The white dwarf star, the core of the star that produced the planetary nebula, is visible in this view of the Ring Nebula, M57, at the center of the doughnut-shaped shell of ejected gas.

44. Open the **Sky Guide** pane, click the **Home** button near the top of the pane (a symbol of a house) and then navigate the following links: **Enter SkyGuide > Night Sky Tours > Twenty great fuzzies > The Ring Nebula.**

Question 23. According to astronomers' estimates, how long ago did the central star of the Ring Nebula blow off this shell of gas?

For high-mass stars, whose mass when on the main sequence exceeded eight solar masses, death arrives in a much more spectacular fashion. These stellar heavyweights end their lives in gigantic thermonuclear conflagrations. In these supernovae, the star's core collapses to become a neutron star or a black hole, while the rest of the star is ejected outward violently, scattering debris into the interstellar medium.

As this debris blasts into the dust and gas of the interstellar medium at supersonic speeds, it excites the surrounding gas and causes it to glow. The visual manifestation of this collision is called a **supernova remnant.**

> Supernova remnants are discussed in Section 20–10 of Freedman, Geller, and Kaufmann, *Universe*, 9th Ed.

In contrast to the gentler ejections that produce the small planetary nebulae around dying stars of lower mass, supernova explosions are so large and violent that the debris hurled into space continues to disperse at supersonic speed for tens of thousands of years. Consequently, many supernova remnants cover fairly large areas of the sky.

45. Select **Favourites > Observing Projects > Milky Way > Crab Nebula.**

46. Use the **Angular Measurement Tool** to find the approximate radius of this object.

M1, the Crab Nebula, is the remnant of a supernova, the light from which arrived at Earth on July 4, 1054 AD. We know this because Chinese astronomers of the time recorded this explosion as a "guest star" that remained visible in daylight for the next 23 days.

Question 24. What is the angular radius of M1?

Question 25. Given that the distance to M1, the Crab Nebula, is 6500 light-years, what is its physical radius in light-years?

Question 26. Given the present size of this nebula and its known age, what will have been its average expansion rate, in kilometers per second? [*Hint:* 1 light-year is equal to about 9.46×10^{12} kilometers, and 1 year contains about 3.17×10^7 seconds.]

H. Distribution of Objects in the Milky Way Galaxy

It is instructive to examine the distribution within the Milky Way Galaxy of the various types of objects that were surveyed in the previous sections of this project.

47. Select **Favourites > Observing Projects > Milky Way > Galactic Object Distribution.**

The view is the same as that of Section A. The southern sky is shown from Cornwall, Canada, at midnight on June 21. However, the horizon, stars, planets, moons, asteroids, and comets have all been removed from the display, leaving only the Milky Way and a line representing the galactic equator.

First, you can examine the location of open clusters in our Galaxy.

48. Open the **Options** pane and expand the layer labeled **Deep Space** and then the **NGC-IC Database** layer.

49. In the **NGC-IC Database** layer, click the **Open Cluster** checkbox to display the positions of many of the open clusters in the Milky Way.

50. Use the **Hand Tool** or cursor keys to survey the sky, noting the distribution of open star clusters in the Milky Way.

Question 27. Relative to the galactic equator, where are most of the open clusters in the Galaxy found?

51. In the **Options** pane, turn off the display of **Open Clusters** in the **NGC-IC Database** layer and select the **Globular Cluster** checkbox instead.

52. Survey the distribution of this sample of globular star clusters in the Milky Way Galaxy.

53. Select **View > Constellations > Astronomical** and **View > Constellations > Labels** from the main menu and repeat your survey of the sky, noting in which constellation or constellations these globular clusters occur most frequently and least frequently.

Question 28. Are globular clusters found in the same distribution as open clusters?

Question 29. Toward which constellation or constellations do most of the globular clusters appear?

In 1920, Harlow Shapley published the results of his observations of RR Lyrae stars in 93 globular clusters. RR Lyrae stars demonstrate a period-luminosity relationship that allows astronomers to determine their absolute magnitude. Comparing the expected absolute magnitude of the star from this relationship with the apparent magnitude as seen from Earth, their distances from Earth can be determined.

Knowing their positions in the sky as well as their distances, Shapley was able to analyze the three-dimensional distribution of globular clusters. He noted that the clusters were distributed in a relatively spherical halo centered over a point lying in the direction of the constellation Sagittarius. Shapley assumed that this distribution surrounded the central gravitational mass of the Galaxy and concluded that the center of the Milky Way Galaxy was approximately 20 kiloparsecs from the Sun in the direction of Sagittarius. Shapley turned out to have overestimated this distance by a factor of about two because he failed to account for the effect of **interstellar extinction,** in which intervening galactic dust artificially reduces an object's apparent brightness and consequently inflates the estimate of its derived distance. Modern observations suggest that the radius of the Galaxy's disk is about 25 kiloparsecs and that the Sun is approximately 8 kiloparsecs from the center.

Question 30. Using the modern observations, what is the approximate diameter of the disk of the Milky Way in light-years?

Question 31. How far from the center of the Galaxy is the solar system in light-years?

I. The Location and Orientation of the Solar System in the Galaxy

54. Select **Favourites > Observing Projects > Milky Way > Overall Structure.**

The view shows a simulated image of the Milky Way Galaxy as it might appear from a distance of 110,000 light-years. Notice the grouping of bright stars centered in the image about two-thirds of the way from the center of the Galaxy, directly below the galactic center. This clump of stars includes the Sun and its nearest neighbors, many of which are the familiar stars that form the constellations as seen from Earth.

55. Use the **Location Scroller** to drag the view upward until the line of sight is along the plane of the disk of the Milky Way (i.e., the Milky Way appears edge-on). If necessary, move the image left or right to spin the image of the Galaxy so that the group of stars representing the Sun's local neighborhood is in a direct line with the center of the Milky Way.

56. Hold down the **Decrease current elevation** button to zoom in on this view until the distance from the Sun as indicated in the **Viewing Location** panel in the toolbar is about **0.2 ly.** Carefully examine the pattern of stars near the Sun and look for a constellation or asterism that lies in the direction of the center of the Milky Way Galaxy. *Hint:* Look for the fish hook asterism to the right of the Sun and the teapot asterism immediately below the image of the Sun.

57. Select **View > Constellations > Labels** and **View > Constellations > Boundaries** from the main menu.

Question 32. Toward which constellations in the Earth's sky is the center of the Milky Way?

58. Select **File > Revert.**

59. Again, use the **Location Scroller** to drag the view so that the plane of the Galaxy is seen edge-on. Then spin the Galaxy so that the center of the Milky Way lies directly between the viewing location and the clump of stars representing the Sun's local neighborhood of stars.

60. Hold down the **Decrease current elevation** button to zoom in on this view until the distance from the Sun as indicated in the **Viewing Location** panel in the toolbar is about **0.2 ly.** Carefully examine the pattern of stars near the Sun and look for a constellation or asterism that lies in the direction of the outer reaches of the Milky Way Galaxy. *Hint:* Look for the belt of Orion in the lower left quadrant of the view.

61. Select **View > Constellations > Labels** and **View > Constellations > Boundaries** from the main menu.

Question 33. Toward which constellation or constellations do you need to look if you want to look at the outer reaches of the Milky Way that lie opposite its center, as seen from Earth?

We can use *Starry Night Enthusiast™* to gain another perspective on the orientation of the plane of the Milky Way relative to the ecliptic that we explored previously.

62. Select **File > Revert** and then use the **Location Scroller** to adjust the view so that the disk of the Milky Way is seen edge-on and is horizontal across the view, with the Sun and its neighboring stars between your location and the center of the Milky Way.

63. Hold down the **Decrease current elevation** button to zoom in on this view until the distance to the Sun displayed in the **Viewing Location** panel is about **2 AU.** In this view, several planets are visible, and the Earth's orbit is shown. Use the **Location Scroller** again to adjust the view slightly so that the ellipse representing the Earth's orbit closes to become approximately a straight line. Then use the **Hand Tool** to adjust the view so that the plane of the Milky Way is once again horizontal across the screen.

Question 34. What is the approximate angle between the ecliptic plane and the galactic plane?

---

64. Select **File > Revert** and take a few minutes to observe the Milky Way from various points of view. Use the **Location Scroller** to view the galaxy face on and look for evidence in this simulated image of dark clouds of gas and dust interspersed among the spiral arms. Note that gas clouds outline the spiral arms and that few individual stars appear within the arms in this view. Note also the position of the Sun and local neighborhood of stars within the Galaxy.

Your observations in the last few steps revealed that the disk of the Galaxy is quite flat. Modern observations indicate that the disk of the Galaxy is only approximately 0.6 kiloparsecs (2000 light-years) thick and the central bulge has a diameter of approximately 2 kiloparsecs (6500 light-years).

J. The Evolution and Spiral Structure of the Galaxy

The disk shape of the Galaxy suggests that the immense cosmic nebula from which the Milky Way formed had some overall rotational component prior to condensing under its own gravity. As the protogalaxy that evolved into the Milky Way condensed out of the foam of the Big Bang, its overall rotation speed increased in order to conserve angular momentum. The increased rotational speed caused material to flatten into a plane perpendicular to the rotation axis of the overall mass.

All of the stars and matter of the Galaxy rotate around its central gravitational mass. The Sun and its system of planets, for example, revolve about the center of the Galaxy with a speed of about 220 km/s and complete one orbit in about 220 million years.

Question 35. Current estimates suggest that the Sun is approximately 5 billion years old. In that time, how many orbits of the Galaxy has it completed?

Question 36. Using a figure of 200,000 years as the age of the human species, through what angle around the center of the Galaxy has the Sun carried us since the dawn of humanity?

Modern radio observations and Doppler studies of the Milky Way Galaxy in the 21-cm radio emission line of neutral hydrogen show that all of the stars, dust and gas in the Galaxy move around its center in the same general direction and with roughly the same speed throughout most of the Galaxy's disk. This is different from the planets, which obey Kepler's laws and move more slowly, the larger their respective orbits. This finding suggests that much of the mass of the Galaxy is contained within a vast spherical halo around its center and takes the form of cold, dark matter.

---

65. Open the **Sky Guide** pane and, from the **Home** page, navigate the following links: Enter **Sky Guide > Our Solar System, the stars and galaxies > Our Galaxy > A closer look at the structure**. Read the information on this page. Then click the link **Just how big is it?** at the bottom of the page and read the next description.

66. Return to the **Home** page in the **Sky Guide** pane (click the button with the house icon) and navigate the following links: Enter **Sky Guide > Night Sky Tours > The voyage of the Chandra Telescope > Milky Way Galaxy > X-ray mosaic of galactic centre** and read the description of the Chandra X-ray image of the center of the Milky Way that appears in the view.

67. Exit *Starry Night Enthusiast*™ without saving changes to the files.

K. Conclusion

In this project, you have examined the Milky Way Galaxy as it appears from Earth and from space. In addition, you have observed many different classes of objects found in our Galaxy and learned about how these various objects evolve.

The Local Neighborhood 22
of Galaxies

Many centuries ago, Greek astronomers debated the question of the position of the Earth within the universe and concluded that the Earth was at the center of it. Later, in the sixteenth and seventeenth centuries, significant advances led to a rationalization of the solar system in which the Sun replaced the Earth as the dominant object, but this Sun was still considered to occupy a central position in the universe. Further observations revealed the true nature of the Milky Way as a galaxy of stars, and the study of clusters of stars within this galaxy led to the realization that the Sun was not even at the center of the Galaxy. Furthermore, observational evidence in the early part of the twentieth century began to show that the Milky Way was not the only large structure of its kind and that the universe was populated by many billions of such structures distributed over vast expanses of space. More recent work has shown that these distant galaxies are assembled into clusters and superclusters of galaxies that form vast walls surrounding relatively empty regions or voids in space. Thus, our increasing base of knowledge has slowly downgraded the status of the Earth from the central object of the universe to an average planet in orbit around a rather ordinary star, accompanying many other stars in a galaxy of stars among billions of such galaxies.

> The Local Group of galaxies is discussed in Section 24–6 of Freedman, Geller, and Kaufmann, *Universe*, 9th Ed.

In this project, you will use *Starry Night Enthusiast™* to explore the Milky Way and its neighboring galaxies, as well as a nearby cluster of galaxies. *Starry Night Enthusiast™* contains a database of the three-dimensional positions of 28,000 galaxies within a distance of about 350 million light-years from the Milky Way, known as the Tully database. Even though this is an enormous distance, the space spanned by this region is a relatively small region of the observable universe, which extends to a distance of about 12 billion light-years and contains billions of galaxies.

A. Zone of Avoidance

As you learned in the project on the Milky Way, our Galaxy contains vast quantities of dust and gas. This material obscures our view of certain regions of the sky from the Earth.

> 1. Launch *Starry Night Enthusiast™*.
>
> 2. Open the **Preferences** dialog window and set the **Cursor Tracking (HUD)** options to **Show: Distance from observer**, **Name**, and **Object type**. Then close the dialog window.
>
> 3. Select **Favourites > Observing Projects > The Local Group of Galaxies > Zone of Avoidance**.

The view is centered on the magnificent spiral structure of the Milky Way, its arms outlined by gas and dust clouds illuminated by bright stars, along with many individual stars dotted along these arms. The arms connect to a rather small central bulge. A bright group of stars near the bottom of the view indicates the Sun's location within a rather indistinct spiral arm.

You can see that many bright points surround the Milky Way Galaxy. Almost all of these are individual galaxies at large distances from your observing position.

> 4. Use the **HUD** feature of the **Hand Tool** to identify a number of these galaxies, noting their distances from your observing location.

You can use the **Location Scroller** to move your viewing location around a sphere centered on the Milky Way. In the upper right corner of the view window, *Starry Night Enthusiast*™ displays the distance of your current viewing location from the Sun and from the object centered in the view, the Milky Way. Moving the **Location Scroller** will change your viewing location around a sphere with a radius equal to your distance from the center of the Milky Way, 146,000 light-years (0.146 Mly) in this case, as shown in the upper right of the view.

> 5. Use the **Location Scroller** to view the Milky Way edge-on, with the Sun directly between the observing location and the center of the Milky Way.

From this perspective, the line of sight is similar to our view from Earth when looking toward the center of the Galaxy. There is a region surrounding the Milky Way where there appear to be no galaxies. This is the **Zone of Avoidance,** where gas and dust in the Milky Way plane have obscured the distant universe from our view on Earth, limiting our knowledge of the distribution of galaxies in these directions. Thus, there might be, and almost certainly are, galaxies there, but we cannot see them from Earth. You can explore the extent of this zone by rolling the Galaxy on its axis.

> 6. Use the **Location Scroller** at one end of the edge-on Galaxy and move it along the plane of the Galaxy's disk to rotate completely around its axis like a wheel. The plane of the Galaxy will tilt as you do this, but you can see the Zone of Avoidance extending across the sky.

You can get another perspective on this zone by moving outward in the Milky Way plane to view the whole Tully database from a distant point.

> 7. Click **File > Revert** to return to the original view, use the **Location Scroller** to move to the galactic plane, from where the Milky Way appears edge-on, and then **Increase current elevation** to view the "rift" in the galaxy field. This is the region where we have been unable to detect galaxies because of the obscuration caused by our own Galaxy.
>
> 8. Use the **Location Scroller** to rotate this collection of galaxies around the Milky Way to observe the three-dimensional nature of this rift.

> **Question 1.** Does the Zone of Avoidance extend all around the sky?
>
> **Question 2.** a) With respect to the center of the Galaxy as seen from Earth, in which direction is the Zone of Avoidance widest? b) In which direction is this zone the narrowest? [*Hint:* Rotate the plane of the Milky Way Galaxy slightly to see where the Sun is.]

B. Galaxy Classification

In this project, you will classify galaxies according to an accepted scheme devised by Edwin Hubble, for whom the highly successful Hubble Space Telescope is named. The classification scheme used in *Starry Night Enthusiast*™ is a more complex variation of the Hubble classification.

Hubble separated galaxies into four main groups according to their appearance:

1) Spirals, which have the designation S

2) Barred spirals, designated SB

3) Ellipticals, denoted by E

4) Irregular galaxies, which have the designation Irr

These galaxies differ not only in appearance but also in composition. Spirals and barred spirals contain lots

> Classification of galaxies is discussed in Section 24–3 of Freedman, Geller, and Kaufmann, *Universe*, 9th Ed.

of dust and gas, and many of their stars appear to be relatively young. Elliptical galaxies contain almost no dust and gas and the component stars appear to belong to an older population. Irregular galaxies contain mostly young stars as well as dust and gas.

A relatively flat disk of stars encircling a central bulge is the distinctive feature of spiral galaxies. The name comes from the apparent spiral distribution of stars and other matter within the disk, beyond the central bulge of these galaxies. Barred spirals are distinguished from other spiral galaxies by a straight bar across the central bulge, the spiral arms in the disk starting at the ends of this bar and not at the edge of the central bulge. Both spiral and barred spiral galaxy groups are sub-classified into three main categories, Sa, Sb, or Sc and SBa, SBb, or SBc, respectively. These sub-classifications are distinguished according to the following criteria:

a) Sa and SBa galaxies have fat central bulges and smooth, broad spiral arms.

b) Sb and SBb galaxies have moderate central bulges and moderately well-defined spiral arms.

c) Sc and SBc galaxies have small central bulges and narrow, well-defined spiral arms.

Differences between Sa, Sb, and Sc galaxies may be related to the amount of dust and gas within their spiral arms. This dust and gas component is important in the production of new stars within galaxies. The approximate percentage of the mass of these galaxies in the form of dust and gas is 4% for Sa galaxies, 8% for Sb galaxies, and 25% for Sc galaxies.

Elliptical galaxies have no disk or spiral arms. Although they are all elliptical in shape, the ellipse can vary from essentially circular to very flat. Hubble classified them in terms of their apparent flattening, from E0 to E7, in which E0 galaxies are the most circular in appearance and E7 galaxies show the most flattening.

The appearance of an elliptical galaxy in this classification scheme does not necessarily describe the true three-dimensional nature of the galaxy. An E0 galaxy might be spherically symmetric or it might be a flattened disk seen face-on from our view. A cigar-shaped galaxy seen end-on might also appear to be spherical and be classified as an E0 galaxy. This is different from the classification of spiral galaxies, where the appearance of the spiral arms indicates the viewing direction on the galaxy from our point of view. We see examples of all directions of alignment of spiral galaxies to our line of sight, so we can be fairly certain that we know the true three-dimensional shapes of these galaxies.

Spiral galaxies contain a great deal of dust and gas and thus appear much more dramatic than elliptical galaxies. In the spiral galaxies, blue and ultraviolet light from young, hot stars illuminates the dust and gas clouds. These stars have been formed fairly recently in the galaxy's history and outline the spiral arms, making them very distinct. In contrast, the lack of dust and gas in elliptical galaxies shows that star formation ceased in these galaxies long ago. Elliptical galaxies are composed of old, red stars and show no evidence of spiral structure. Nevertheless, elliptical galaxies come in the widest range of sizes and masses, from the largest known galaxies to very small aggregates of old stars. Often, the gravity from the mass of one or two giant elliptical galaxies will dominate the motions of entire groups or clusters of galaxies.

> The tuning-fork diagram is shown in Figure 24–11 of Freedman, Geller, and Kaufmann, *Universe*, 9th Ed.

Hubble summarized his classification scheme by drawing the types of galaxies on a tuning-fork diagram in which the handle contains the elliptical galaxies, from E0 at the end to E7 near the forks. The spiral and barred spiral galaxies occupy the two tines of the fork, arranged from Sa and SBa near the handle to Sc and SBc at the ends of the tines.

At the junction of the handle and the tines, Hubble placed another group of galaxies he called lenticular galaxies. These resemble elliptical galaxies in that they lack gas and dust and show no spiral structure, but they appear to have a definite central bulge. Hubble denoted these as S0 for normal lenticular galaxies and SB0 galaxies if they showed a bar.

The irregular galaxies did not fit into this tidy scheme, but Hubble defined two types: 1) Irr I galaxies, which look like undeveloped spiral galaxies and contain many young stars and lots of dust and gas, and 2) Irr II galaxies, which are distorted and show no particular symmetry. They appear to have been formed by collisions with other galaxies or disturbed by violent activity within their interiors.

9. Select **Observing Projects > The Local Group of Galaxies > Milky Way** from the **Favourites** pane. Then close the **Favourites** pane.

The view, which has been adjusted to show only the nearest galaxies in order to reduce the confusion on the screen, is a representation of our Galaxy, the Milky Way. It shows a magnificent spiral structure, with arms outlined by gas and dust clouds illuminated by bright stars, along with many individual stars dotted along these arms. These distinct arms connect to a rather small central bulge. The Sun's position, indicated by the label, is within the bright group of the local stars that make up the familiar constellations.

10. [Optional] To see the Milky Way against the background of the more distant galaxies in the Tully database, select **Options > Deep Space > Tully 3D Database...** and in the dialog window, adjust the **Visibility range** slide control all the way to the right (**Far**). You can then click the **Cancel** button to remove the distant galaxies again.

11. Use the **Location Scroller** to view the Milky Way from different perspectives including edge-on.

12. Use the **Angular Measurement Tool** to determine the proportion of the disk that is occupied by the central bulge.

Question 3. On the basis of the appearance of the Milky Way Galaxy from above its spiral arm plane, what would be the Hubble classification of this galaxy?

Question 4. Expressed as a percentage, what proportion of the Milky Way's diameter does the central bulge occupy?

C. The Milky Way and its Closest Neighbors

In this section, you will use *Starry Night Enthusiast*™ to investigate the size and classification of several of the galaxies that are companions to the Milky Way.

13. Select **File > Revert.**

14. Open the **Object Contextual Menu** for the **Milky Way** and select **Show Info.**

15. Note the **Galaxy thickness** and **Galaxy diameter** of the Milky Way under the **Other Data** layer of the **Info** pane. Copy these values to Data Table 1 below.

Question 5. Expressed as a percentage, what is the ratio of the maximum thickness of the Milky Way Galaxy to its diameter?

16. Use the **Location Scroller** to adjust the view so that the disk of the Milky Way Galaxy is seen edge-on and horizontal, with the Sun between the viewing location and the center of the Milky Way.

17. **Increase current elevation** to about **0.150 Mly** from the Sun.

You will notice two somewhat diffuse and nebulous regions in the lower right quadrant of the view.

18. Identify these two objects with the **HUD**.

You will see that these objects are the Magellanic Clouds, prominent naked eye objects in the skies of Earth's southern hemisphere. The Small Magellanic Cloud appears blue. This is the result of light from hot, young stars within this galaxy being scattered and reflected by the dust within it. The Large Magellanic Cloud shows a redder color.

In the next few steps, you can zoom in to take a closer look at each galaxy before classifying them.

19. Open the **Object Contextual Menu** over the **Small Magellanic Cloud** (SMC) and select **Show Info**. From the **Other Data** layer of the **Info** pane, copy the values for the **Galaxy thickness** and **Galaxy diameter** for the Small Magellanic Cloud into Data Table 1 below.

20. Open the **Object Contextual Menu** over the **Small Magellanic Cloud** again and select **Center**. Note that the distance data shown in the upper right corner of the main view window now shows the distance of the observing location from the Sun and from the Small Magellanic Cloud.

21. **Decrease current elevation** until the distance from the Small Magellanic Cloud, shown in the upper right corner of the view, is about **15,000 ly** (light-years) away.

22. The Small Magellanic Cloud almost fills the view from this distance. Use the **Location Scroller** to observe this companion galaxy to the Milky Way from various points of view, including edge-on and face-on.

Question 6. What is the Hubble classification of the Small Magellanic Cloud?

Question 7. Is there any evidence of spiral structure or central bulge in the Small Magellanic Cloud?

Question 8. Expressed as a percentage, what is the ratio of the thickness of the Small Magellanic Cloud to its diameter, as given in the **Other Data** layer of the **Info** pane?

23. Select **File > Revert** and notice that the Large Magellanic Cloud is visible in the view to the right of the Milky Way. Open the **Object Contextual Menu** over the **Large Magellanic Cloud** and select **Show Info**. From the **Other Data** layer of the **Info** pane, copy the values for the **Galaxy thickness** and **Galaxy diameter** for the Large Magellanic Cloud into Data Table 1 below.

24. Open the **Object Contextual Menu** over the **Large Magellanic Cloud** once more and select **Centre**.

25. **Decrease current elevation** until your distance from the Large Magellanic Cloud is about **40,000 ly** and use the **Location Scroller** to view the Large Magellanic Cloud face-on and edge-on to examine its structure.

Question 9. What is the Hubble classification of the Large Magellanic Cloud?

Question 10. Is there any evidence of spiral structure or central bulge in the Large Magellanic Cloud?

Question 11. Expressed as a percentage, what is the ratio of the thickness of the Large Magellanic Cloud to its diameter, as given in the **Other Data** layer of the **Info** pane?

26. Select **File > Revert** and use the **Location Scroller** so that the Milky Way is edge-on. Then use the **Location Scroller** to roll the Milky Way along its disk until the Sun is on the opposite side of the center of the Milky Way from the viewing location.

Your eye might be caught by a small image that appears just below the Milky Way plane. Let us examine this fascinating object in more detail.

27. Move the **Hand Tool** over this small diffuse object and identify it as **Sagittarius** from the **HUD**.
28. Open the **Object Contextual Menu** for **Sagittarius** and select **Centre**.
29. **Decrease current elevation** until your distance from **Sagittarius** is about 8000 ly.
30. Use the **Location Scroller** to look at this galaxy from different perspectives.
31. Open the **Object Contextual Menu** for **Sagittarius** and select **Show Info**. From the **Other Data** layer of the **Info** pane, copy the values for the **Galaxy thickness** and **Galaxy diameter** for Sagittarius into Data Table 1 below.

As you see, this galaxy is approximately symmetrical in all axes and appears to have no dust or gas within it. These observations should allow you to place this galaxy within the Hubble classification scheme. This type of galaxy is certainly *very* different in character from both the Milky Way Galaxy and the Magellanic Clouds.

Question 12. What is the Hubble classification of Sagittarius?

Data Table 1. Physical Data for the Milky Way and its Nearby Companion Galaxies

Galaxy	Diameter (ly)	Thickness (ly)
Milky Way		
Small Magellanic Cloud		
Large Magellanic Cloud		
Sagittarius		

Question 13. How much larger is the diameter of the Milky Way compared to the diameter of a) the Small Magellanic Cloud, b) the Large Magellanic Cloud, and c) Sagittarius? [*Hint:* Divide the diameter of the Milky Way by the diameter of each galaxy.]

D. The Distribution of Galaxies Near the Milky Way

A further characteristic of galaxies in our universe is their distribution in space. These massive collections of stars and other matter also seem to congregate together in clusters. The number of galaxies in a cluster can vary widely. Those with few members are known as poor clusters and are often termed groups, while those with many hundreds and thousands of galaxies are known as rich clusters. Furthermore, their distribution around some center of concentration leads to a second classification. Those clusters with a distinct spherical appearance are called regular clusters, whereas those whose galaxies appear to be scattered randomly over a large region of space are called irregular clusters.

The properties of these massive conglomerates of galaxies are not yet fully understood. For example, gravitational effects have been observed in galaxies and clusters of galaxies that appear to require far more mass than is provided by the obvious visible matter, resulting in the postulation of **dark matter** that cannot be observed directly by present technology.

We find that our Milky Way Galaxy is a member of a poor cluster known as the Local Group. This group contains at least 40 other galaxies and extends out to a distance of approximately 10 million light-years from the Milky Way. Faint new members of this Local Group are still being discovered as astronomers develop improved observing techniques. Many members may never be discovered because they lie within the Zone of Avoidance, obscured by the gas and dust of the Milky Way.

The two largest subgroups within the Local Group are Group 223, of which the Milky Way is the dominant member, and Group 222, dominated by the Andromeda Galaxy.

32. Select **Observing Projects > The Local Group of Galaxies > Milky Way and Magellanic Clouds** from the **Favourites** pane.

The view shows the Milky Way and the Magellanic Clouds and illustrates the three-dimensional relationship between them.

33. Use the **Angular Measurement Tool** to find the distance from the Milky Way to each of the Magellanic Clouds and to Sagittarius. [*Note:* This distance is displayed in blue, below the angular separation measurement.] Record these distances in Data Table 2 below.

34. Select **File > Revert.**

35. **Increase current elevation** to about 600,000 light-years from the Sun (**0.600 Mly**).

36. Use the **Location Scroller** to move the view back and forth.

In the previous step, you may have noticed that, in addition to the labeled galaxies, several other of the points of light on the screen appear to have relatively larger motion than others and appear to move along with the Milky Way. Those points of light that show this phenomenon are other small galaxies that, along with the galaxies labeled in the view, form a subgroup of the Local Group named Group 223. The dominant galaxy of this subgroup is the Milky Way.

37. Open the **Object Contextual Menu** over the **Milky Way** and select **Highlight "223" Group.**

38. **Increase current elevation** to about 2 Mly to see the extent of this subgroup of galaxies within the Local Group.

In the next sequence, you will make observations of the member galaxies of this subgroup.

39. Select **Observing Projects > The Local Group of Galaxies > Group 223** from the **Favourites** pane.

The view is from about 2 million light-years away from the Milky Way, which is centered in the view. Member galaxies of Group 223, which includes the Milky Way, are highlighted in yellow. The view is restricted to only the closest galaxies. The thousands of background galaxies are artificially hidden from the view.

40. [Optional] Select **Options > Deep Space > Tully 3D Database...** set the **Visibility range** slide control in the dialog window all the way to the right to see the background galaxies in the view. Then click the **Cancel** button.

41. For each of the highlighted galaxies in the view, select **Magnify** from the **Object Contextual Menu** and use the **Location Scroller** to examine the galaxy you have chosen and to determine its Hubble classification. Then select **Show Info** from the **Object Contextual Menu** and note the dimensions of this galaxy from the **Other Data** layer of the **Info** pane. Use Data Table 2 below to keep track of this information. Then select **File > Revert** and repeat this step for another highlighted galaxy. [*Note:* You do not have to repeat these observations for the Magellanic Clouds, Sagittarius or the Milky Way since you have obtained the data for these galaxies of Group 223 in the previous section.]

Data Table 2. Hubble Classification and Physical Properties of Member Galaxies of Group 223

Galaxy	Hubble Classification	Galaxy Diameter (ly)	Galaxy Thickness (ly)
Sculptor Dwarf			
Fornax			
Carina			
Sextans			
Ursa Minor Dwarf			
UGC 10822/Dr			
Leo I			
Leo II			

Use the data you have compiled in Data Tables 1 and 2 above to answer the questions that follow.

> **Question 14.** Which is the largest member of the subgroup of galaxies called Group 223?

> **Question 15.** To which Hubble class do most of the galaxies in this group belong?

42. Select **File > Revert.**
43. Select **Options > Deep Space > Tully 3D Database...** and click the checkboxes labeled **Entire dataset** and **Highlighted filaments/groups** in the dialog window that pops up. Then click the **OK** button.

Starry Night Enthusiast™ has outlined the extent of Group 223 against the extent of the Tully database. Recall that in this view the background galaxies in the Tully database are not shown.

44. Use the **Location Scroller** to observe this group from various perspectives. Adjust the viewpoint so that the box bounding Group 223 is seen face-on. Use the **Angular Measurement Tool** to measure the angular distance between the Milky Way and the two opposite sides of the long axis of the box. Add these angular distances, rounding to the nearest degree, and record the result in Data Table 3 below. Do the same for the width of the bounding box around Group 223. Then adjust the view with the **Location Scroller** so that the bounding box is edge on and measure its angular height in a similar fashion.

45. Using a value of 2 million light-years as the distance of your viewing location from the Milky Way, you can determine the physical dimensions, in light-years, of the extent of this group of galaxies dominated by the Milky Way by multiplying your angular measurement in degrees by 2 million light-years divided by the number of degrees in one radian, which is 57.3°. The resulting calculation reduces to:

$$\text{Physical Distance (ly)} = \text{Angular Distance (°)} \times 34904$$

Data Table 3. Dimensions of Group 223 from a Distance of Two Million Light-years

Measurement	Angular Distance (°)	Physical Distance (ly)
Length of Bounding Box		
Width of Bounding Box		
Thickness of Bounding Box		

Question 16. What is the approximate volume of space occupied by Group 223?

Question 17. What is the result of the previous question when expressed in units of cubic kiloparsecs? [*Hint:* Recall that a parsec = 3.26 light-years and that a kiloparsec = 1000 parsecs.]

The dimensions of even our tiny region of the universe are staggering. Light, traveling at the fastest possible speed in our universe, takes tens of thousands of years to reach us, even from our nearest neighbor galaxies. It takes hundreds of thousands and even millions of years to travel from the more distant members of our small subgroup of galaxies. And yet, these galaxies are close compared to the distant field of galaxies in the Tully database included in *Starry Night Enthusiast*™, which extends as far as 350 Mly from us. Our own subgroup of galaxies is certainly in our backyard compared to the most distant galaxies recently imaged by the Hubble Space Telescope and by several large new telescopes in Hawaii and South America. These modern instruments have produced images from light that left these much more distant galaxies up to12 billion years ago and has only recently reached Earth!

E. The Andromeda Subgroup of Galaxies

While the Milky Way dominates the Group 223 subgroup contained in the Local Group of galaxies, the Andromeda Galaxy, which is visible to the naked eye from Earth under dark-sky conditions, dominates the other major subgroup of galaxies in the Local Group.

46. Select **File > Revert.**

47. **Increase current elevation** to about **4 Mly** from the Sun.

As you increase the distance from the Sun, the Andromeda Galaxy comes into the view from the right.

48. Identify the Andromeda galaxy with the **HUD.**

49. Open the **Object Contextual Menu** for this galaxy and select **Show Info.** Note the dimensions of this galaxy from the **Other Data** layer in the **Info** pane.

50. Open the **Object Contextual Menu** for the Andromeda Galaxy again and select **Magnify.**

51. The distance of the viewing location from the Andromeda Galaxy is shown in the upper right corner of the window. Use the **Decrease current elevation** button to set this distance at about 140,000 light-years (**0.140 Mly**).

52. Use the **Location Scroller** to examine this galaxy and its nearby neighbors.

Question 18. Is the Andromeda Galaxy larger or smaller than the Milky Way?

Question 19. What is the Hubble classification of the Andromeda Galaxy?

Question 20. Does the Andromeda Galaxy, like the Milky Way, appear to have close companions? If so, identify them using the HUD.

The Andromeda Galaxy is the dominant galaxy in Group 222, the other major subgroup of galaxies in the Local Group.

53. Select **Observing Projects > The Local Group of Galaxies > Group 222** from the **Favourites** pane.

The view is from a location about 1.3 million light-years from the Andromeda Galaxy, which is centered in the gaze.

54. Open the **Object Contextual Menu** for the **Andromeda Galaxy** and select **Highlight "222" Group.**

55. **Increase current elevation** until the distance to the Andromeda Galaxy as shown in the upper right corner of the view window is about **10 Mly.**

56. Select **Options > Deep Space > Tully 3D Database...** and click the checkboxes labeled **Entire dataset** and **Highlighted filaments/groups** at the top of the dialog window that appears. Then click the **OK** button.

57. Use the **Location Scroller** and **Elevation** controls in the toolbar to examine this subgroup of galaxies.

Question 21. Which is the richer subgroup of galaxies in the Local Group?

Question 22. Which is the larger of the subgroups in the Local Group?

58. Select **Observing Projects > The Local Group of Galaxies > Local Group** from the **Favourites** pane.

The view is from a position in space over 6 million light-years from the Sun, with the Milky Way centered in the gaze. The galaxies within the local group, including those from both of the subgroups you have examined in the previous sequences, are highlighted.

59. Use the **Location Scroller** to look at the Local Group of galaxies from different points of view.

60. Select **Options > Deep Space > Tully 3D Database...** and click the checkboxes labeled **Entire dataset** and **Highlighted filaments/groups** at the top of the dialog window. Then click the **OK** button.

61. Use the **Location Scroller** to view the boundaries of the Local Group.

Question 23. Would you describe the Local Group as a rich or a poor cluster?

Question 24. Would you describe the Local Group as a regular or irregular cluster?

F. The Location of the Local Group within the Universe

After this brief look at a few of the Local Group of galaxies, you can conclude this exploration by using *Starry Night Enthusiast* ™ to show you where this cluster of galaxies fits into the larger scale of things.

62. Select **File > Revert**.

63. Select **Options > Deep Space > Tully 3D Database...**

64. Move the dialog window that appears to the upper left corner of the screen and watch the view as you gradually move the **Visibility range** slide control in the dialog box toward the right until it is about at the midpoint of the scale. Click the **OK** button to exit the dialog.

65. Use the **Location Scroller** to look around the view, now populated with more distant clusters of galaxies. As you shift your viewpoint, notice the Zone of Avoidance. Adjust the view so that the widest part of the Zone of Avoidance is in the direction of the Milky Way, centered in the view.

66. Gradually **Increase current elevation** to about **100 Mly** from the Sun. You will notice a rich cluster of galaxies come into the view. This is the Virgo cluster of galaxies. To highlight its member galaxies, open the **Find** pane and type **The Eyes**, the name of one of the member galaxies of the Virgo cluster, in the search box and press the **Enter** key.

67. The view centers on the Virgo cluster. Click the menu button in the **Find** pane next to **The Eyes** listing and select **Highlight "GA Virgo Cluster" Filament.**

Question 25. Does this cluster appear richer or poorer than the Local Group?

68. Select **Observing Projects > The Local Group of Galaxies > Virgo Cluster** from the **Favourites** pane.

The view is from a location within the Virgo cluster of galaxies and the gaze is centered on the galaxy labeled "The Eyes. "

69. Use the **Location Scroller** to look around the view from within this rich cluster of galaxies.

70. **Zoom Out** to a maximum field of view.

71. Gradually **Increase current elevation** to about **70 Mly** from the Sun. The Milky Way, labeled, comes into the view.

72. Position the **Hand Tool** so that the HUD indicates **The Eyes**. Activate the **Angular Measurement Tool** to find the distance between The Eyes and the Milky Way, ensuring that the HUD indicates that you are pointing to the Milky Way at the conclusion of the measurement.

Question 26. How long does it take light from the stars in the galaxy called The Eyes to reach the Earth?

73. Continue to **Increase current elevation** gradually, pausing to look around the view with the **Location Scroller** at intervals.

As you venture farther into space, you will find that galaxies tend to be found in rather thin walls. These walls surround vast voids in which few, if any, galaxies are found. Clusters of galaxies are found within these walls and the whole structure of deep space seems to resemble that of a collection of soap bubbles. The walls of galaxies and the more concentrated clusters and superclusters are found at the interstices of these bubbles. You will have the opportunity to explore this larger scale structure of the universe in the next project.

G. Conclusion

You have explored the relationships between our Galaxy and its near neighbors in the Local Group of galaxies and measured the scale of this space. You have also classified many members of this loose, poor cluster of galaxies. Finally, you observed a rich cluster of galaxies relatively near to the Milky Way and you were then able to place yourself within the large-scale structure of the universe in a way that two-dimensional pictures could never do, allowing you to glimpse the bubble-like features that lace our local extragalactic neighborhood.

Large-Scale Structure of the Universe **23**

Our own Milky Way Galaxy is a gravitationally bound structure consisting of a huge, flattened disk made up of about 200 billion stars and large quantities of dust and gas in spiral arms surrounding a central bulge. Our increasingly precise and penetrating observations of the universe over the past century have revealed that the Milky Way is but one of many billions of such collections of matter, many of them part of immense clusters and superclusters of galaxies. The last

> Section 24–6 in Freedman, Geller, and Kaufmann, *Universe*, 9th Ed. discusses clusters, super-clusters, and the "foamy" structure of the universe.

few decades in particular have led us to the realization that even galaxies outside these huge galaxy clusters are not randomly distributed in space but occur within huge walls surrounding vast voids that contain few, if any, galaxies. Thus, the present view of the large-scale structure of the universe is one that resembles a collection of soap bubbles. Clusters and superclusters of galaxies appear to occupy the lines where the walls of the bubbles (or voids) meet, the largest of them being where two or more lines meet. Furthermore, at least one of these superclusters, dubbed the Great Attractor, appears to exert sufficient gravitational influence on galaxies, including our own Milky Way, that they are moving collectively toward this region of space.

In this project, you will use *Starry Night Enthusiast*™ to venture among these vast structures, stopping at important way stations to examine the detailed and interrelated collections of galaxies that make up our corner of the universe.

A. The Milky Way and the Local Wall

You can start your journey from the Milky Way Galaxy and move slowly outward to visualize its position within the Local Group, a poor cluster of more than 40 galaxies.

> 1. Launch *Starry Night Enthusiast*™.
> 2. Select **Favourites > Observing Projects > Large-Scale Structure of Universe > Milky Way**.

The view shows the Milky Way from a location 110,000 ly from the Sun. Before speeding off into deep space, it is instructive to examine the universe from this "local" position. In particular, there is a significant region on our sky that is blocked from our view because matter within the Milky Way absorbs the light of distant galaxies, and it is helpful to look at the effect that this **Zone of Avoidance** has on our view.

> 3. Use the **Location Scroller** to move your view of the Milky Way so that it appears edge-on.

Almost all of the bright points in the view are distant galaxies, but you will note the apparent absence of such galaxies around the plane of the Milky Way. This lack of galaxies is not real but shows our ignorance of this region of the sky because of the absorption of light by material in our own galactic environment. You should keep this limitation in mind as you explore further into space.

4. Use the **Location Scroller** to roll the Milky Way around its axis like a wheel to demonstrate that the Zone of Avoidance extends completely around our position, as expected. Also notice that this Zone of Avoidance is larger in extent when the view is toward the center of the Galaxy when compared to the view out of the Galaxy. The plane of the Milky Way will tilt during this movement, but you can still see this apparent "rift" in the distribution of galaxies in line with this plane.

As you move your viewpoint, you will notice that local galaxies such as the Large and Small Magellanic Clouds and the giant Andromeda Galaxy move relatively rapidly across the view.

> **Question 1.** What is the term for the effect of closer galaxies appearing to shift their positions against the more distant background galaxies as your point of view changes?

You can take a brief look at this group of neighboring galaxies.

5. Select **File > Revert** and then use the Location Scroller to place the Milky Way edge-on again, with its plane horizontal across the view.
6. To reduce the confusion from distant galaxies, select **Options > Deep Space > Tully 3D Database...** and in the dialog box, move the slider on the **Visibility range** scale so that the left edge of the slider button aligns with the **Near** tick mark.
7. [Optional] You may want to adjust the **Brightness** control toward the **More** setting.
8. Click the **Labels** check box in the **Tully 3D Database Options** dialog box to turn this option on.
9. Click the **OK** button to exit the **Tully 3D Database Options** dialog and return to the main view.
10. **Increase current elevation** to about **3.5 Mly** from the Sun.

You will see the Milky Way and its near neighbor galaxies recede into the distance as the other sub-cluster of galaxies around the giant Andromeda Galaxy comes into view at about this distance, in the bottom-left corner of the view.

11. **Increase current elevation** further to about **7.5 Mly** from the Sun.

A few of the outlying galaxies of our Local Group come into the view.

12. Gradually **Increase current elevation** to about **30 Mly** from the Sun, where you will see many more galaxies move into the view.
13. Select **Labels > Tully 3D Database** to turn off the labels.
14. Use the **Location Scroller** in a top-left to bottom-right motion to rotate around the position of the Milky Way in the center of the view window, to see how the Local Group of galaxies fits into a narrow wall of galaxies. You can use the **Location Scroller** to move this wall to an edge-on position and then roll it along like a wheel to explore its relationship with other structures in this region of space.

This wall, inside which the Milky Way resides, separates two voids, as shown by the lack of local galaxies in these regions.

15. Position the cursor over the Milky Way and open the **Object Contextual Menu**. Select **Highlight "GA Coma-Sculptor Cloud" Filament** from the **Object Contextual Menu** to highlight this wall of galaxies.

16. Use the **Location Scroller** to look around the view and then align the gaze along the long axis of this narrow wall of galaxies. To see how this structure got its name, select **View > Constellations > Labels**.

17. Select **Options > Deep Space > Tully 3D Database...** and click the checkboxes at the top of the dialog window labeled **Entire dataset** and **Highlighted filaments/groups**. Then click **OK** to close the dialog window.

18. **Increase current elevation** to about 85 Mly from the Sun and use the **Location Scroller** to look around the view again.

Starry Night Enthusiast™ displays a cube representing the extent of the entire Tully database of 28,000 galaxies. Within this cube is a box representing the boundaries of the highlighted GA Coma-Sculptor Cloud. The GA designation indicates that this wall is part of the Great Attractor supercluster.

Question 2. Toward which constellations does the long axis of this wall of galaxies that contains the Milky Way point?

19. To view this wall of galaxies edge-on, use the **Location Scroller** to rotate the view so that the Milky Way is in the direction of the label for the constellation Circinus.

20. Use the **Angular Measurement Tool** to obtain an estimate of the thickness of this wall of galaxies. To do so, measure the angular distance from the Milky Way in a direction parallel to the edge of the bounding box facing the constellation Tucana, to about the midpoint of the side of the wall facing the label for the constellation Musca. Add this result, to the nearest degree, to the measurement of the angular distance from the Milky Way to the midpoint of the side of the wall that is facing the label for the constellation Norma. Record the total, to the nearest degree, in Data Table 1 under Thickness.

21. Use the **Location Scroller** to view this wall face-on so that the Milky Way is aligned with constellation label of Canis Major. Use the method employed in the previous step to estimate the length and height, or width, of this wall of galaxies and record the results in Data Table 1.

Data Table 1. Dimensions of the GA Coma-Sculptor Cloud

Dimension	Angular Size (°)	Angular Size (Radians)	Dimension Size (Mly)
Thickness			
Length			
Width			

Question 3. What are the approximate angular dimensions of this structure?

22. To determine the size of the dimensions of this wall in millions of light-years, first convert your angular size measurements in degrees to angular size in radians. To do this, divide the angular size in degrees by 57.3, the number of degrees in 1 radian. Enter the results in the column labeled **Angular Size (Radians)** in Data Table 1.

23. Now, multiply each of the angular sizes in radians by the distance of the observing location from the Sun, which is shown in the **Viewing Location** pane of the toolbar.

Question 4. What are the approximate dimensions of this wall of galaxies in millions of light-years (Mly)?

Question 5. What is the approximate volume of this wall of galaxies in cubic light-years?

It is perhaps chastening to remember that each bright point in this image is a galaxy in its own right, each one containing many thousands to millions of stars, and that you are looking (in simulation, admittedly) at the universe on a truly vast scale.

24. Select **Labels > Constellations** to turn off these labels.

25. Use the **Location Scroller** to explore this region of the universe from different viewpoints.

Moving around the Sun's position at this distance, you get the sensation of moving through bubbles of space with thin walls of galaxies separating large voids. You can move the sky around to show that the wall that you have highlighted is connected to several superclusters of galaxies at the interstices of these bubbles.

26. Select **Labels > Constellations** to turn this feature back on.

27. Use the **Location Scroller** to position the Milky Way over the label for the constellation Coma Berenices.

In this view, you will see a tight cluster of numerous galaxies to the left of the Milky Way.

28. Position the **Hand Tool** over this cluster of galaxies and open the **Object Contextual Menu**. Select **Highlight "GA Virgo Cluster" Filament**. Again, you can use the **Location Scroller** to see the relationship between the wall containing the Milky Way and this cluster, in the corner of a void.

29. Use the **Location Scroller** to position the Milky Way at one of the vertices of the bounding box surrounding the Virgo Cluster of galaxies and use the **Angular Measurement Tool** to estimate the angular sizes of one of the sides of the bounding box surrounding this cluster. Assuming that the cluster is spherical, this measurement approximates the angular diameter of this cluster of galaxies.

30. Position the **Hand Tool** over several of the highlighted galaxies in this cluster (you may want to **Zoom In** on the view to do this) and use the data from the **HUD** to obtain an estimate of the distance of this cluster from your observing location.

Question 6. What is the approximate diameter of this cluster of galaxies in millions of light-years (Mly)?

Question 7. What is the approximate volume of this cluster in cubic light-years?

Question 8. From your observations and calculations, which of the following is correct?

    a) The Virgo Cluster occupies a smaller volume in space than the Coma-Sculptor wall that contains the Milky Way.

    b) The Virgo Cluster appears to contain many more galaxies than the Coma-Sculptor Wall.

    c) The Virgo Cluster is a much richer cluster of galaxies than the Coma-Sculptor Wall.

    d) All of the above are correct.

31. To see the whole of a giant wall of galaxies that includes the Virgo Cluster and the Coma-Sculptor Cloud that contains the Milky Way, select **Favourites > Observing Projects > Large-Scale Structure of Universe > Supergalactic Equatorial.**

The view is zoomed in on the Milky Way from a distance of about 80 Mly from the Sun.

32. **Zoom Out** to the maximum field of view to see the highlighted galaxies that form this gigantic wall.

33. Use the **Location Scroller** to rotate the sky again to demonstrate that this is a relatively flat wall by finding the direction from which this structure is edge-on to your viewpoint and note that it lies along the line drawn on the screen representing the extragalactic equator. This wall is used to define an "Equatorial Plane" for reference when discussing this region of extragalactic space.

34. **Increase current elevation** to about **200 Mly** from the Sun. Use the **Location Scroller** to roll this wall like a wheel to see that the wall has linear structures within it and that narrow subsidiary walls are attached to this vast wall and spread away from it in all directions like the branches of a Christmas tree.

35. Position the **Hand Tool** over the any one of the galaxies in the dense cloud of galaxies near the center of this wall and open the object's contextual menu. The checked **Highlight** item in the menu will identify the cluster of galaxies to which this galaxy belongs.

Question 9. Which cluster is near the center of the Supergalactic Equatorial Wall of galaxies?

B. Superclusters of Galaxies

As you saw at the end of the last section, large superclusters of galaxies lie at the ends of walls of galaxies. You can examine one of these superclusters in more detail with the following procedure.

36. Select **Observing Projects > Large-Scale Structure of Universe > Virgo Cluster** from the **Favourites** pane.

The view is centered on the Milky Way from a location 100 Mly from the Sun. At this position, you can easily see the wall that you explored above. The Virgo Cluster is highlighted in the upper right of the view. As you have seen, this is a rich cluster of galaxies containing about 3000 member galaxies.

37. Use the **Angular Measurement Tool** to measure the distance from the Milky Way to M87, the labeled galaxy at the heart of the Virgo Cluster. The distance between these objects is displayed beneath the angular separation measurement.

Question 10. How far is the Virgo Cluster from the Earth?

38. Select **Options > Deep Space > Tully 3D Database...** and slide the **Visibility range** control all the way to the left (**Near**); then click **OK**.
39. Position the **Hand Tool** over the Virgo Cluster, and select **Centre** from the **Object Contextual Menu**.
40. **Decrease current elevation** to about **70 Mly** from the Sun.

The galaxy M87 is labeled in the view.

41. Position the **Hand Tool** over M87 and select **Centre** from the **Object Contextual Menu**.
42. Use the **Location Scroller** to look all around the Virgo Cluster, noting the position of M87 within it.

Question 11. What is the position of M87 within the Virgo Cluster?

43. Open the **Object Contextual Menu** over M87 and select **Go There**.
44. Use the **Location Scroller** to look at this galaxy from different viewpoints.

Question 12. What is the Hubble classification of M87?

This galaxy is very bright at radio wavelengths and is also known by its radio source name of Virgo A. The galaxy contains a prominent jet emanating from a bright star-like core. It is obvious that some very energetic processes are going on in this core to produce this energy and the jet.

45. Click the **Home** button in the toolbar.
46. Open the **Find** pane.
47. Click the magnifying glass icon at the left side of the search box in the **Find** pane and select **Chandra images** from the menu. Then type **M87** in the search edit box.
48. Click the menu button for the Chandra M87 Image in the list and select **Magnify**. If *Starry Night Enthusiast*™ informs you that this object is not currently visible, click the **Best Time** button and then click the menu button for the M87 Image in the find pane and select **Magnify** once more. The view shows an X-ray image of this galaxy obtained by the Chandra X-ray Observatory.
49. To view the equivalent optical image of this very active galaxy, open the **Options** pane and expand the **Deep Space** layer. You can reduce and even remove the Chandra image from this composite image by using the slider bar to the right of the Chandra heading to see the optical image of the elliptical galaxy.

You can also look briefly at the enigmatic jet of material emanating from the core of this galaxy. *Starry Night Enthusiast*™ has a Hubble Space Telescope image of this jet, somewhat offset from M87 to avoid confusion.

50. **Zoom Out** to a field of view of about **34** arcminutes. The Hubble image is just to the upper right of M87.

51. Move the cursor over this image, open the **Object Contextual Menu** and center the view on this image, known as the Cosmic Searchlight. **Zoom In** to a field of view of about **50** arcseconds to see this enormous jet of material being emitted by some mechanism from the core of the galaxy.

Obviously, at least some giant elliptical galaxies have gigantic powerhouses at their cores, probably associated with supermassive black holes.

C. The Great Attractor

The galaxies in a large volume in the vicinity of the Milky Way appear to be moving toward a region of the sky filled with a relatively high density of galaxies that make up a complex structure. Because of this coordinated motion, this region toward which these galaxies appear to be moving has been called the Great Attractor. We can travel out to a distance from which we can view this as an entity, examine its structure and position, and measure its distance from the Milky Way.

52. Open the **Favourites** pane and select **Observing Projects > Large-Scale Structure of Universe > Great Attractor.**

The view is centered on the Milky Way from a position about 300 Mly from the Sun. The highlighted galaxies comprise the Great Attractor, a huge pinwheel structure almost filling the view with a giant supercluster at its center. Many thousands of galaxies make up this structure, and the attractive force that this structure exerts on the surrounding galaxies shows that there is a high concentration of mass there.

53. Use the **Location Scroller** to move the Great Attractor around to see its three-dimensional structure and locate its approximate center.

54. You will notice the large, rich cluster at the center of the Great Attractor. Position the **Hand Tool** over this cluster and open the **Object Contextual Menu.** The checked **Highlight** option indicates the name of this collection of galaxies.

Question 13. Which cluster lies at the center of the Great Attractor?

In rotating the Great Attractor around in the sky, you will note that it has a long, linear feature. You can measure the linear extent of this overall structure.

55. **Increase current elevation** to about **500 Mly** from the Sun and use the **Location Scroller** to rotate the view so that the long axis of the Great Attractor stretches across the screen. Then use the **Angular Measurement Tool** to measure the distance between two highlighted galaxies at either end of this axis.

Question 14. What is the approximate length of the linear structure of the Great Attractor?

Note that this linear structure contains the Virgo Supercluster and that there is an almost continuous line of galaxies from end to end along this linear structure.

D. The Great Wall and other Colossal Walls

In this section, you will explore several other huge wall-like structures of galaxies in the extensive region of the universe surrounding our position in the Milky Way. From distances that take you to the limit of the present Tully Collection of galaxies, you can look in on these structures and discover the beautiful bubble-like shapes within the walls of which these galaxies reside.

> 56. Open the **Favourites** pane and select **Observing Projects > Large-Scale Structure of Universe > Great Wall.**

The view is from about 500 Mly from the Sun and the structure called the Great Wall is highlighted.

> 57. Use the **Location Scroller** to examine this vast structure from various points of view. Adjust the view so that you are looking at this structure edge-on. Again, this wall is thin and relatively flat and extends over a huge distance.

This vast collection of galaxies is arranged on a surface surrounding a huge volume almost devoid of galaxies.

> 58. Open the **Favourites** pane.
>
> 59. Select **File > Revert** and, after observing this view of the Great Wall for a moment, select **Observing Projects > Large-Scale Structure of the Universe > Great Wall and Great Attractor** from the **Favourites** pane.

The view shows both of the structures you examined in the previous sequences: the Great Wall and the Great Attractor. Notice that the region between these two vast structures contains relatively few galaxies.

Random processes cannot have formed this type of structure and such structures provide an important clue to the evolution of our universe following its origins in the Big Bang. There are other equally impressive structures stretching across this corner of the universe that have become apparent as astronomers have determined distances to galaxies and placed them in this three-dimensional mesh.

> 60. From the **Favourites** pane select **Observing Projects > Large-Scale Structure of Universe > Southern Wall.**

The view shows another huge curtain of galaxies, somewhat closer to the Milky Way and smaller than the Great Wall and on the opposite side of the sky from our viewpoint.

> 61. Use the **Location Scroller** to examine this structure from different points of view.
>
> 62. From the **Favourites** pane, select **Observing Projects > Large-Scale Structure of Universe > Two Walls.**

The view highlights both the Southern Wall and the Great Wall. In addition to the Milky Way, two other galaxies are labeled in the view: ESO 605-16 in the Southern Wall and MCG 3-31-2 in the Great Wall.

63. Use the **Angular Measurement Tool** to find the distance between these two labeled galaxies and between these two galaxies and the Milky Way.

Question 15. What is the approximate distance between the Great Wall and the Southern Wall as measured by the distance between the two representative galaxies labeled in the view?

Question 16. From your measurement of the position of the Milky Way with respect to these two walls of galaxies, which of the following do you think is correct?

    a) The Earth and the Milky Way are at the center of the universe.

    b) The Tully database is derived from Earth-bound observations and encompasses only a small region of the universe around the Milky Way.

    c) Both A and B are correct.

    d) Neither A nor B are correct.

In this section, you will explore one of the regions of the nearby universe that has a relatively sparse number of galaxies surrounded by walls of richer clusters.

64. Open the **Favourites** pane and under **Observing Projects > Large-Scale Structure of Universe**, select the following views in turn: **South Pole Void - front, South Pole Void – back.**

The two rich lanes of galaxies you observed in the last step surround a region that has relatively few galaxies within its volume.

65. Select **Favourites > Observing Projects > Large-Scale Structure of Universe > South Pole Void – all.**

66. Use the **Location Scroller** to examine this void from different viewpoints.

E. Beyond The Tully Collection

The Tully Collection of galaxies extends out to about 600 Mly and *Starry Night Enthusiast*™ provides you with the opportunity to examine the type and position of the full 28,000 galaxies in three-dimensional space. This remarkable data set is nevertheless limited and, whereas astronomers have a wide range of information about more distant objects, conclusions about these more distant realms are more speculative.

In this wide-ranging project, you have explored the large-scale structure of our universe in a way that is impossible in real life, using some of the best data in the world to travel vicariously though the soap-bubble features of the realm of the distant galaxies. You can investigate individual galaxies and collections and their characteristics and spacing at will by following the guidelines and methods outlined in this project.